Living Zen Remindfully

Also by James H. Austin

Zen-Brain Horizons (2014)
Meditating Selflessly (2011)
Selfless Insight (2009)
Zen-Brain Reflections (2006)
Chance, Chase, and Creativity (2003)
Zen and the Brain (1998)

Living Zen Remindfully

Retraining Subconscious Awareness

James H. Austin, M.D.

The MIT Press
Cambridge, Massachusetts
London, England

This book was set in Palatino and Frutiger by Toppan Best-set Premedia Limited. Printed and bound in the United States of America.

Library of Congress Cataloging-in-Publication Data

Names: Austin, James H., 1925– author.
Title: Living Zen remindfully : retraining subconscious awareness / James H. Austin, M.D.
Description: Cambridge, MA : MIT Press, 2016. | Includes bibliographical references and index.
Identifiers: LCCN 2016015015 | ISBN 9780262035088 (hardcover : alk. paper)
Subjects: LCSH: Meditation—Zen Buddhism. | Awareness—Religious aspects—Zen Buddhism. | Consciousness—Religious aspects—Zen Buddhism. | Zen Buddhism—Psychology.
Classification: LCC BQ9288 .A935 2016 | DDC 294.3/4435—dc23 LC record available at https://lccn.loc.gov/2016015015

10 9 8 7 6 5 4 3 2 1

In memory of Scott Whiting Austin (1953–2014)
To my early teachers Nanrei Kobori-Roshi, Myokyo-ni, and Robert Aitken-Roshi for their inspiration; and to countless others whose contributions to Zen, to Buddhism, and to the brain sciences are reviewed in these pages

The Zen Way is a demanding way, but it
leads to the depths, to the light of clearly
seeing what is when the veil is rent, and
to the warmth of the heart that touches
and engenders growth.

Myokyo-ni (1921–2007)[1]

The only true law is that which
leads to freedom. There is no other.

Jonathan Livingston Seagull[2]

Contents

Chapters Containing Testable Hypotheses

List of Figures and Tables

Preface

To some persons nowadays, mindfulness might seem to be just another short course. After auditing it for only a few weeks, they could thereafter meditate casually, whenever …

This isn't where *Living Zen Remindfully* is coming from. Authentic Zen training means committing oneself to a process of regular, ongoing daily life practice. This preparation enables one to *unlearn* old unfruitful habits, retrain more wholesome ones, and lead a more genuinely creative life.

Currently, mindfulness usually means that one is consciously aware of events in the immediate *Now*, in "the *present* moment." Then why will the word, "*sub*conscious" appear so often in these pages? A short answer here leads us back to Sigmund Freud. (A longer explanation is presented elsewhere.[1-6]) Freud considered that our mind's conscious compartment was only the small visible tip of a large iceberg. Barely "one-seventh of its bulk" could be seen floating above water. Therefore, most of our brain's activities—those other six-sevenths—were submerged, out of sight. Similar comparisons captured public attention after an iceberg sank the Titanic in 1912. Then, eight-ninths or even nine-tenths became common estimates for how much lurked in our subconscious mind. There, its hidden potentials often seemed readier to access our worst behavior, not our best.

This book extends some original implications of mindfulness. It explores the many positive, helpful subconscious aspects of *remindfulness*. It suggests ways that long-term meditative retraining can help cultivate hidden, affirmative resources of our subconscious memories. Accessing these subtle processes of transformation can enable us to adapt more effectively and to live more authentic lives.

To these goals, part I reviews different types of meditation and briefly discusses what *enlightenment* means. It then considers how long-term meditative retraining influences creativity and sponsors the development of character.

Part II addresses the fundamental biological distinction: Self/other. We're reminded that certain nerve cells in the medial temporal lobe are already sensitive not only to such contrasts as being Self-centered (*ego*centric) or being other-centered (*allo*centric) but also to the different ways we blend our focal and global attention.

Part III reviews the remarkable processes that first encode our memories, then store them, and later retrieve them. It focuses more on those covert, helpful *remindful* processes that have been incubating problems at subconscious levels, less on situations that we remain all too Self-consciously aware of.

Part IV takes off from there. It considers why ancient aspects of Zen, thorny issues in clinical neurology, and concepts at the cusp of neuroscience research have already become topics that are mutually illuminating.

Part V illustrates what "Living Zen" means in the everyday life of any human being. This person could be communing with Nature outdoors, meditating indoors on a cushion, living an active or quiet life anywhere. One example of this creative principle in the daily life of a Zen practitioner will be that of Basho, the seventeenth-century master of *haiku* poetry.

Don't be surprised when you encounter the different topics in parts I—V set next to each other in unconventional ways. The short essays in the appendices also provide brief updates on related issues.

For the reader's convenience, bracketed references in the text indicate specific pages in earlier books in this series. For example, when you see [ZB], the numbers refer to useful background information in *Zen and the Brain*. [ZBR] refers to *Zen-Brain Reflections*; [SI] to *Selfless Insight*; [MS] to *Meditating Selflessly*; and [ZBH] to *Zen-Brain Horizons*.

Chapters and appendices that contain testable hypotheses are listed on page xii.

Acknowledgments

I'm indebted to Philip Laughlin at the MIT Press for appreciating the need to bring this volume to the attention of the wider meditating and neuroscience communities. Again, my sincere thanks go out to Katherine Arnoldi Almeida for her skilled editorial assistance and to Yasuyo Iguchi for her artistic skill in designing the cover, icons, and figures.

I am especially grateful to Barbara Klund for her ongoing dedication in deciphering my handwriting throughout multiple drafts of her excellent typing and for helping to keep this manuscript organized as it expands. Many thanks go also to James W. Austin for his valued assistance, together with Seido Ray Ronci, in reviewing and commenting on the manuscript.

Janice Gaston is my invaluable muse and close consultant on Audubon, musical, and related issues.

In recent years I have been privileged to share in the inestimable bounties of regular Zen practice with our sangha at Hokoku-an led by Seido Ray Ronci, with the Vipassana/Theravada activities of the local Show-Me Dharma sangha, with visits to the Live Oak Sangha, and in the stimulus afforded by the annual Mind and Life Summer Research Institute. Gassho to all!

While this book manuscript was underway for the MIT Press, an article on Basho was accepted by the *Eastern Buddhist*, to appear in a subsequent issue under the title, "A Plop! Heard 'Round the World."

By Way of a Personal Introduction

This is the sixth book of words about Zen. My interest in Japanese culture began back in 1950. A Navy reservist, I was called up and assigned to be the neurologist at an Army hospital in Yokohama. Almost a quarter of a century later, my given scientific reason for returning to Japan was to conduct a sabbatical research project in neuropharmacology. It happened that the best place to do this was in Kyoto!

By fortunate links in an unforeseen chain of circumstances, this led to my starting to meditate at Daitoku-ji, a Rinzai Zen temple. My teacher there was an English-speaking Zen master, Nanrei Kobori-Roshi. This became the start of a long interdisciplinary quest to understand the relationships between Zen and the brain.

This book serves as an update on the recent progress along the rapidly expanding, exciting interface between Zen and the neurosciences. I still write from the perspective of a student—of biology, of neurology, and of Zen—awed by the challenges involved in attempting to integrate ancient wisdom with the stimulating new research findings that are arriving every day.

Part I

On the Path of Meditation

When one goes into Zen meditation,
one passes, as a usual process,
through a psychic field,
from the surface down into the depths,
as if one were plummeting
into a lake
in a
diving
bell.

Nanrei Kobori-Roshi (1914–1992)

Can Meditation Enhance Creative Problem-Solving Skills? A Progress Report

> Buddhism has always maintained that the best place to practice Buddhism is precisely *where you are*, and that the environment that needs tending is the *interior* environment.
>
> Charles Prebish[1]

> A state of conscious awareness resulting from living in the moment is not sufficient for creativity to come about. To be creative, you need to have, or be trained in, the ability to carefully observe, notice, or attend to phenomena that pass your mind's eye.
>
> M. Baas and colleagues [2]

My first interest was in creativity. Long before I meandered into Zen, I was curious about the role that insight played in the creative process.[3] Zen, insight, and creativity are three huge topics. It helps to start with some basic dictionary definitions:

- *Creativity*: In general, the word refers to the long and complex series of interactions between an individual and the environment that culminate in something *new*.

- *Intuition*: The generic faculty of direct knowing. Intuition usually refers to our ordinary levels of intuitive understanding, to hunches that proceed quickly without obvious rational thought processes.

- *Insight*: The sudden act of seeing clearly and comprehensively—with *no* intervening thoughts—into problem situations or into our own inner nature. Insight implies a more refined level of processing than intuition.

- *Insight-Wisdom*: The profound, insightful comprehension that realizes the existential essence of all things.

- *Kensho* (J.): A brief, extraordinary alternate state when consciousness sees deeply into the existential essence of things. In Zen Buddhism, the state of *kensho* (and the next more deeply awakened state, termed *satori*) are viewed as only the beginning of authentic spiritual training.

- *Prajna* (Skt.): The flashing process of enlightenment that awakens the state of insight-wisdom.

- *Zen*: A form of Mahayana Buddhism that attempts to express the wordless physiology of this direct, selfless awakening to the whole natural world.

Subsequent publications explored these links among the topics of creativity, meditation, and Zen.[4] Readers will discover that there is now an international interest in how these three large topics illuminate each other. Our goal in this first chapter is to sample recent research. Accordingly, the next section begins with creativity. Later sections relate creativity to meditation. In part II, we'll consider the central topic of Self and how meditation helps dissolve our maladaptive Self-centeredness. Meanwhile, don't get hung up on the neuroanatomy. Just follow the narrative as it unfolds.

Creative Problem Solving Using Matchsticks

Jing Luo and colleagues in Beijing first reported their pioneering fMRI research on insight processing back in 2004 and 2007. [SI: 158–159, 169] The latest study from this group[5] builds on the ways we first need to decompose big "chunks" of data before going on to solve the kinds of problems that are posed in the "matchstick test." [SI: 183–184] By way of explanation, this test uses Roman numerals that are constructed of matches (as shown in the icon of part I). One example of this kind of problem might be illustrated below in the following manner:

Start with this *false* equation, composed of matches,

$$VI = VII + I$$

Convert these Roman numerals into a true statement by moving only one match.

The solution:

$$VII = VI + I$$

Why must we remain *flexible* throughout such a task? First, in order to achieve *Destructuring*. This means breaking up prior rigid expectations, fixed perceptual patterns, big unwieldy chunks of data. After this, the big intractable problems can become smaller, more *separable elements*. These elements are now more mobile and capable of rejoining in different combinations. We also require flexibility, almost simultaneously, in order to achieve *Restructuring*. This means reorganizing these fluid ingredients into a novel, meaningful, *appropriate* solution. Moreover, all during these multiple sequences, we must keep *disregarding* irrelevant information.

In their latest experiment, the subjects were 18 Chinese undergraduate or graduate students.[6] All were from the Beijing University of Science and Technology. These students had a twofold task based on ideograms. It was much more complex than the matchstick test.

1. First, they needed to transform the small parts of one Chinese character into those of another Chinese character.

2. Then they had to specify whether this second version was a "real," familiar character, one that was appropriate for daily use in real life, and not a "pseudocharacter."

Did the students use approaches that were novel and appropriate? Functional MRI monitoring during this *dual,* originating/judgmental task, revealed the following correlations:

- *Novelty processing*: Novelty processing correlated with *increased* neural activity in three regions: the *caudate nucleus* (L > R) (considered here to be an index of procedural memory); the *substantia nigra* (mental energy/rewarding); and the standard visual-spatial processing regions. *Decreased* neural activity occurred in the cuneus, lateral temporal gyrus (Brodmann area [BA] 20, 21, 22), inferior parietal lobule (BA 39, 40), insula (left), medial prefrontal cortex, and precuneus. [SI: 60, 61]

- *Appropriateness processing*: This evaluation stage correlated with *increased* activity. It was referable to the inferior parietal lobule (L > R), lateral temporal gyrus (BA 20, 21) (L > R), left superior frontal gyrus (BA 8), left medial frontal gyrus (BA 10), left anterior and posterior cingulate (BA 32, 31), and precuneus.

What happened when the students processed "chunks" that were not only *novel* but also *appropriate*? Notice that these activations were referable to the hippocampus (considered to be an index of declarative memory) and amygdala (emotional arousal). Both of these regions are in the medial temporal lobe (chapter 6).

These ideogram tests involved both spatial and linguistic recognitions of standard versus novel word symbols. The authors concluded that *both* forms of memory—the nondeclarative kinds as well as the kinds that could be declared in words—contributed jointly to the two key defining features of creative cognitive processing: its novelty and its appropriateness. In part III, we continue to discuss these various aspects of memory and consider how meditation influences them.

Divergent Approaches to Creativity

How did Beaty and colleagues first screen their 91 subjects for creativity skills?[7] They tested them with a battery of six computerized tasks. These tasks were oriented toward *divergent* (non-focused) styles of thinking. At that point they selected two well-matched groups: the top 12 subjects formed the high-creative group; the bottom 12 were the low-creative group. They then assessed the functional connectivity of these two groups using functional MRI (fMRI) during the so-called "resting state." To do this, they selected the two seed-based frontal lobe regions of the *ventral* attentive processing system: the right and left inferior frontal gyrus (BA 47). [SI: 32–33, 232]

The data confirmed that divergent styles of creative performance draw upon certain parts of *both* hemispheres. The high-creative group members showed greater connectivity linking their *left* (*not* right) inferior frontal gyrus (IFG) with the entire so-called default mode network. In this network major contributors were the medial prefrontal cortex and the posterior cingulate cortex.* These same high-creative group members also showed greater functional connectivity that linked their *right* inferior frontal gyrus, their inferior parietal cortex bilaterally, and their *left* dorsolateral prefrontal cortex (DLPFC).

So, greater divergent thinking performance did correlate with an increase in functional connectivity between the *left* inferior prefrontal cortex and the default-mode network.

---

* This network maintains a high metabolic rate. However, unlike the old basal metabolic rate (BMR), these brain regions are not actually "resting" during most conditions under which they are being measured. This introduces other complexities. [SI: 70–76]

The evidence that the *right* IFG was more connected with the *opposite*, left dorsolateral prefrontal cortex was intriguing for two reasons: (1) this left DLPFC was a region being linked with our top-down control of attention and with our capacities for working memory; (2) the authors speculated that one of the default network's roles in sponsoring mind wandering might actually help support the dynamic range of one's random, useful "imaginative processes." The authors also considered the possibility that creativity would require some higher executive control of the conceptual combinations arising from within this network. Why? Because we subsequently need to critique the various options. This critical overview judgment clearly differs from our earlier impulses to simply generate numerous ideas. [ZBH: 159]

Oriental Gamesmanship

It's been obvious for centuries: a few experts play certain board games much more skillfully than do the rest of us novices. Faster than a finger snap, South Korean experts in the game of GO categorize whole, complex scenes. Instantly, they deploy the kinds of perceptual templates that account for their fast processing skills. Each subconscious *gist* confirms their implicit, memory-based expectations. Where do some diffusion-weighted MRI results suggest that connections in white matter are also involved? In two subcortical regions: beneath the right frontal and left inferior temporal cortex. [ZBH: 161–162]

Three comprehensive reports from Japan have now compared experts against novices while each played a Japanese style of chess called *shogi*. The standard form of this game is much more complicated than the way we play chess in the West. In the first study from the Riken Brain Science Institute, the fMRI responses of 11 professional players were contrasted with the responses of 17 amateur players (8 high-ranking; 9 low-ranking).[8] Each day, the *shogi* professionals had usually been concentrating on their competitive games for three to four hours.

The *posterior precuneus* became activated when these professionals *first* perceived realistic patterns on the *shogi* board. Next, as they quickly generated their best move, they then activated their *caudate* nucleus (R > L). The authors noted two ways that projections from this precuneus could go on to reach the head of the caudate: (1) through the relay that would first pass through the dorsolateral prefrontal cortex and (2) by a direct route to the head of the caudate in its dorsomedial portion.

In their next report[9] the authors added a longitudinal training component to this first study of game-playing intuition. For 15 weeks, they now trained 20 male novices (aged 20–22 years) to play competitive *mini-shoji*. This mini version is simplified. Its board has only 25 squares, not 81. Each player has only 6 pieces to move, not 20. Substantial monetary payments helped to motivate the best players.

A challenging checkmate situation then confronted every player. They had only *two* seconds to decide which move to make. This quick-next-move response was used to measure their intuitive skills. Importantly, it was again the head of their right *caudate* nucleus (not other cortical regions) that continued to develop more activation during this 15-week period of training. Greater degrees of this right caudate activation also paralleled their increasing percentages of correct responses. However, the *volume* of this right caudate head did not change during this short course of study. (Nor did the volume of the caudate nucleus of the professionals in their first [2011] study differ from that of the high-level amateurs.)

What makes these results noteworthy is that the problem-solving creativity tasks that involve actual *movements* in space (matches, chess pieces) seem to be activating the caudate—a higher-level *motor* nucleus in the basal ganglia.

In the latest fMRI study,[10] 17 high-ranked *shoji* players faced a more strategic quick decision: Given where all these pieces are now positioned on the standard (large) *shogi* board, is my best move to attack? Or should I play defense? Notice how the Self and its "turf" is now becoming involved. Emotional orientations—outward versus inward—are infusing this seemingly "cognitive" task. Under *these* Self-othering experimental conditions, the optimum strategic values actually out on the board (but not the strategy each player actually selected) were correlated with the players' fMRI activities at *either* end of their *cingulate* gyrus. For example, the optimum *defensive* strategy values (not their actual selections) correlated with their *rostral anterior cingulate* activations. The optimum *offensive* strategy values correlated with their *posterior cingulate* activations. Moreover, different activities in the dorsolateral prefrontal cortex appeared to reflect the differences intermediate between these two competing, emotionally-intrusive, "strategic" evaluations.

These *shoji* samurai tended to attack. They were prone to respond in this outward-aggressive manner. This bias was an overinvestment in what they each guessed would be a favorable benefit/risk ratio. However, neither their heart rates nor amygdala activities nor ventral striatum activities necessarily correlated with the decisions that they actually made.

Commentary

Creativity is multifaceted. Its different skill sets are reflected in the ways these individual subject groups differ, and in the ways the task conditions differ among different experimenters. A broad, novel generalization begins to emerge in the comprehensive review of creativity by Jung et al.[11] The authors consider a theory that "creative cognition" often expresses a "disinhibitory bias" in brain networks, which otherwise "normally stand in excitatory and inhibitory balance." [ZBH: 87–88] (appendices B and D comment further on the importance of such inhibitory mechanisms and on those released during disinhibition.)

In the recent large college student cohort (n = 766) studied by Takeuchi et al.,[12] there appeared a cascade of linked associations. The tendency for greater creativity to be expressed in divergent thinking, especially in women, correlated with these factors: the dopamine D_2 receptor gene (an association further enhanced during stress), emotional intelligence, and the state of motivation.

The next issue seems straightforward: Can meditators become more intuitive in practical ways?

Does Meditation Enhance Creative Problem-Solving Skills?

Psychologists at the University of Amsterdam approached this big question several ways.[13] They enlisted naive undergraduates (n = 450 or so). Next, they estimated—using *self-reports*—these students' attention/awareness skills. The estimates were based on the subjects' responses to the Mindful Attention Awareness Scale (MAAS) and to the Kentucky Inventory of Mindfulness Skills scale (KIMS). The students were then trained to focus on *observation* skills during an 8-week course coupled with brief tape-guided periods of meditation. Thereafter, various idea-generation tests assessed the students' actual mindful and creative skills. The data from four sets of experiments were interpreted in support of two main conclusions:

- *Observational* skills (when coupled with cognitive *flexibility*) reliably predicted not only creative achievements but also creative behavior and fluency. This evidence supports a crucial point suggested elsewhere: to be creative, our brain needs to make skillful, *flexible* shifts at *precisely the right times*.[14] [ZB: 258; ZBH: 166]

• Simply deploying awareness, per se, was not linked with creativity. This evidence could support multiple reports elsewhere that "mindfulness" contains multiple components, each having different mechanisms of action. For example, the *receptive* meditation approach used in Colzato's 2012 study[15] enhanced the subjects' *divergent* styles of thinking when they performed the Alternate Uses Task. The tendency for the *concentrative* meditative approach to improve *convergent* thinking was less obvious. [ZBH: 156–157, 185]

In their 2014 report, Baas et al. also reviewed the existing research on different styles of meditation.[16] The authors included this caveat: "Research on meditation is still in its infancy." There exists a serious "lack of studies that systematically distinguish between and [that] compare different kinds of meditation on various cognitive, affective, and executive control tasks."

With this caveat in mind, it becomes useful at this point to briefly summarize, in table 1.1, some major aspects of the three categories of meditation that Lippelt et al. cited in their recent review.[17]

The contemporary literature cited in this review article reveals a serious limitation: a dearth of *explicit* information that is focused specifically on the longer, completely *thought-free intervals* that arise in long-term meditation practitioners. Why might one hope that this reflects only a lack of familiarity with the major "no-thinking" orientations of Zen practice?[18] The fact is that brief, calm, clear, spatially aware intervals with no word thoughts can also arise relatively early along the meditative Path. [ZB: 68] Moreover, some profound creative potentials can open up when one's incessant internal monologue drops off, and absolutely *no*

Table 1.1
Types of Meditation

Basic Type of Meditation	Basic Description	Comment	Usual Implications	Prior Discussions
Concentrative Meditation (Focused Attention Meditation)*	Intention chooses deliberately to focus one's attention on an external source or on the phases of one's breathing.	This is a short-term memory task. It already involves a more-or-less conscious, subtle monitoring* of one's attentive focus.	Is useful as a way to train attention in general, to increase the stability, clarity, and span of one's attention to the present moment.	ZBR: 33–35, 37 SI: 27–29 MS: 63–69, 104, 203–204
Receptive Meditation (Openly Receptive Meditation)	After intention lets go of the prior narrow focus of attention, one enters into a calm, clear, open awareness that simply notices internal and external events.	This is an openly receptive, increasingly bare awareness. Subconscious surveillance mechanisms spontaneously notice when monkey-mind-wandering strays too far and return it automatically.	Is useful as a way to train awareness in general and to support long-term, covert intuitive capacities.	ZBR: 36, 38 SI: 29–39 MS: 72–74, 92–103, 104, 203–204
Lovingkindness Meditation Practice	One's focus of visual and verbal attention is on developing kindness and compassion—first toward oneself. These positive emotional resonances are then extended toward all other beings.	This is a consciously directed practice. Often, it is introduced near the close of a period of meditation.	Is useful as a reminder to become more kind to oneself and to all others during daily life activities off the cushion.	ZBR: 48–50, 270 SI: 41–42, 146–147

*The meditator's short-term memory capacities are constantly being used to monitor the success of this initial focusing. Therefore, it seems both inaccurate and confusing to try to insert the word "monitoring" in the separate title used to describe the openly receptive type of meditation. [ZBR: 30; MS: 167, 203–204] *word-thoughts* are heard. [SI: 76, 150–158; MS: 84, 146, 165; ZBH: 81–83, 147–151, 155–156] *This silence is golden.* [ZB: 633–636] It comprehends.

Short-Term Meditation

Authentic, eyes-open Zen training retains its old-fashioned orientation—toward the very *long*-term transformations of character that we discuss in chapter 3. However, short term, 5-day approaches have been used to instruct undergraduates at a technical University in China. [SI: 39–43] This approach is called "integrative body-mind training" (IBMT). It confirms that our top-down skills of *convergent* attentive focusing are relatively easy for beginners to learn. In contrast, advanced meditators are slower to develop their range of more mature, *divergent* "opening up" meditative styles. These allow them gradually to access the much more subtle range of *subconscious* receptive practices that tap into hidden resources of focal and global information processing (chapters 10 and 18).

Two recent reports of short-term results were also based on a similar undergraduate cohort at Dalian University in China. In the first,[19] 20 undergraduates completed this IBMT program. It lasted 30 minutes a day for only 7 days. The 20 student controls received the same amount and duration of muscle relaxation training (RT). Both groups averaged 21 years of age. The subjects' mood during the previous week was assessed using the Positive and Negative Affect Schedule (PANAS). The short Torrance Tests of Creative Thinking (TTCT) were used to assess the kinds of diversity that index the fluency, flexibility, and originality of responses. In brief, the IBMT Group did increase their Torrance diversity scores more than the relaxation group ($p < 0.01$). Further analyses suggested that the subjects' increase in positive moods and their decrease in negative moods could have infused some of these beneficial, affective limbic contributions into their short-term increases in creativity. [SI: 176; ZBH: 158] We're reminded that the ancient Greeks had coined the word *muse* when they referred to the ways their nine goddesses, all sisters, had inspired them to rise to new heights of creativity in music, poetry, the arts, and science.

The second report[20] assessed the important issue of insight (at least to the degree that our insight's ordinary short-term problem-solving aspects are currently assessed with the aid of a Remote Associations Test [RAT]). [SI: 158–160; 170–173] The researchers began with these working hypotheses: (1) the *cingulate* gyrus could be active in both detecting and resolving a cognitive conflict; (2) the *insula* could be active during the awareness of an error, a useful attribute when one is starting to adjust to any such error; (3) other necessary ingredients that included meaning and comprehension could be supplied during the whole process by the *inferior parietal lobule* and the *superior temporal gyrus*.

The 16 undergraduates in the body-mind meditation group were trained for 30 minutes each day—this time for 10 days. The relaxation group received the same hours and duration of muscle relaxation. Each group of trainees then saw three different Chinese words. Their task was to think of a single, fourth solution word that formed a complete, coherent phrase. An extra cue might be supplied, lasting two seconds, to help mobilize a latent solution that could resolve the earlier impasse. These RAT sequences were monitored with fMRI. The researchers calculated that using a "time window of 6–11 seconds, after the presentation of the standard solution," would allow them "to isolate the corresponding hemodynamic response."

The two groups showed comparable fMRI results *before* training. However, after training, the IBMT students showed significantly greater activations in their *right* cingulate gyrus, insula, putamen, and inferior frontal gyrus. This IBMT group also showed greater activations *bilaterally* in the middle frontal gyrus (L > R), inferior parietal lobule (R > L), and superior temporal gyrus (L > R). These several activations could help confirm that the ten-day IBMT group was more successfully involved in "highly integrated processes such as conflict detection, breaking mental set, restructuring problem representation, error detection, and the experience of 'Aha!' in a moment."

Recent Experiments in Ordinary Problem Solving: Contrasts between Ordinary Insight and a More Analytical Approach

The so-called "Aha!" experience is a *post*-insight phase. It correlates with the arrival of happiness and related positive emotions.[21]

Salvi and colleagues[22] measured their subjects' overt visual attention during three distinct intervals: (1) preparation; (2) presentation of the problem word associations, and (3) the moment just prior to the solution. During the preparation interval, those college students who would later solve the problems by insight blinked more frequently, blinked longer, showed fewer eye movements, and *looked outside the problem box* more often than did those students who would only solve the problem analytically. The authors suggested that these blink rates were related to phasic increases in dopamine functions. They also concluded that the above differences in performance were driven by the subjects' "starting and unfolding" their states of attention. "Unfolding" serves as a reminder: *attention*—in some form—*needs to be ongoing all during attentive processing.* [ZBR: 179–183; ZBH: 191]

Zedelius and Schooler[23] followed up on the premise that an increased tendency to mind wander would imply a lesser tendency to be mindful *at that moment*. They recently studied the behavioral responses of older, educated subjects (average age 36 years) to compound remote word association tasks. Mindfulness was estimated by self-reports. The effect of meditation was not assessed. Higher mindfulness scores in both experiments correlated with *decreased* problem solving overall. On the other hand, suppose the subjects used an *analytical* strategy to solve the remote association tasks (either a strategy that the subjects had usually employed spontaneously or one that was now being directed by instruction). Now their subjects' greater degrees of mindfulness were consistently linked with an *increase* in problem-solving performance.

The reader who has arrived at the gist of these recent reports can now understand why we need to consider many networking connectivities and modules in order to help interpret the multiple mechanisms and sequences that enter into different kinds of creative processing.

In Conclusion

So the short answer is *yes*, to the question: Can meditation enhance creative problem-solving skills? However, it remains to be determined which major modules are consistently involved, and into which precise sequences their diverse functions are inserted. Moreover, most experiments have focused on *ordinary* insights. These experiments depended on tests that involved word associations or ideograms. This research on ordinary insight omits a significant question: why does a relative absence of the sense of Self occur during a major moment of insight? Furthermore, which millisecond sequences are involved during the acute state of insight-wisdom in *kensho-satori*?

A complete survey of creativity is far beyond the scope of this slender volume, but several recent articles sample the current issues now expanding this big topic.[24] A similar recent expansion has also occurred in the large field of meditation research.[25] A special issue (October 2015) of the *American Psychologist* surveys the many critical questions about mindfulness and meditation that illustrate why "Research in this area is still in its infancy."[26] A recent follow-up study of previous results is sobering.[27] Ninety-seven of the original 100 articles had earlier reported statistically significant results. However, these results (first published in three general psychology journals) were later replicable in only 36 instances!

Given this preamble on insight, in the next chapter we ask: What does the word *enlightenment* mean? What are the implications of such an "awakened" state of consciousness?

2

In Zen, What Does It Mean "To Be Enlightened"?

If you investigate and inquire diligently for a long time with single-minded concentration, the time of fruition will come. ... Suddenly, as the bottom drops out of the bucket, you will empty out and awaken to enlightenment.

Ch'an Master Yuanwu Keqin (1063–1135)[1]

Enlightenment means to directly see the essential world through one's experience. But to bring the enlightened eye to complete clarity requires a long period of continued practice.

Koun Yamada-Roshi (1906–1989)[2]

Enlightenment is a word that means different things to different people in different cultural traditions.[3] To briefly survey Zen Buddhist awakening with some sense of structure, it helps to begin by separating the several brief *states* of insight-wisdom from a later *stage*. A useful classification is to place in one large category three major *advanced*, extraordinary *states* of sudden enlightenment.[4] This classification clearly lumps together these three progressively deeper *states* called (1) *kensho*, (2) *satori*, and (3) Ultimate Being. [ZB: 303, table 1; 51–624; 625–636] Such an approach distinguishes these three *states'* transient but transforming phenomena from that later, rarer, ongoing *stage* called the Exceptional *Stage* of Ongoing Enlightened Traits. [ZB: 636–695] *That* stage is what Yamada-Roshi would aspire to. Indeed, he continued to practice a Living Zen for decades after his major awakening at the age of 47.

This general formulation, separating states from stages, resembles a system of successive mental levels described by Zen Master Hakuin Ekaku (1689–1769).[5] Hakuin was exceptionally qualified to speak about the wide range of insights. Like master Tahui six centuries earlier, he could have had as many as 18 major awakenings and innumerable little insights. [SI: 154][6]

Among Hakuin's multiple insights of various sizes was one that he described as his decisive enlightenment experience.[7] It occurred one spring night in 1726. He had paused and was reading the Lotus Sutra. In the silence, a cricket suddenly issued a series of chirps. Instantly, all his doubts dissolved: "A loud involuntary cry burst from the depths of my being, and I began sobbing uncontrollably." From that pivotal moment on, I "lived in state of great emancipation." Later chapters and notes[8] will keep reminding us about this decisive state's three salient points of interest: its auditory trigger, its sudden release into the realization of total freedom, and, in Hakuin's case, his authoritative conclusion that its transforming sense of liberation was not only *complete* but also *ongoing*. [ZB: 580–581, 638–641]

Hakuin's approach to enlightenment also started from his premise that there existed (at least) two levels or degrees in these *states* of insight-wisdom.

- The first of these advanced levels realized what might be translated into English as a "universal cognition of equality." This penetrating insight into the nature of reality realized that all sensations and knowledge were expressions of one's essential enlightened (Buddha) nature.* Hakuin cited one personal example of what the world could look like at that moment: it carried the unmistakable impression that it was *one's own image*.[9] This experience of coidentity was like "looking at your [own] face in a mirror." [ZBR: 344–345]

- Hakuin's next level was a state describable in English as consistent with a "subtle observing cognition." This advanced degree of insightful comprehension discerned what "Reality" *really* was—its underlying experiential principles and continuum of differences. The result would become a strong commitment to skillful action, to actually helping everyone else. Notice that a profound sense of authentic compassion is implied in this description. It suggests how deeply etched are the transformations of character that can follow a major state of *satori*.

- Hakuin's third level was a *stage*. This was the ongoing *stage* of "practical cognition." In its fully developed form, this rare stage implied that all of the person's ongoing actions were finally the *living* embodiment of genuine *wisdom*, living in full accord with the nature of reality. The individual was now completely liberated, having dropped off all prior unfruitful overconditioning. In this stage, virtuous behaviors now flowed effortlessly. How effortlessly? Like "the actual fruit forming after the flower petals have dropped away." [ZB: 668–677]

A meditative practice need not have conventional "spiritual" overtones. Single, transient states of "enlightened

---

* We are a tiny integral part of what many might think of as Mother Nature, The Great Spirit, etc.

awakening" are relatively common events in the population at large. [ZBR: 6–7; ZBH: 221–222, n 11] The different brains of different people experience awakenings differently. At least eighteen descriptions of *kensho's* insight-wisdom can be identified. [ZB: 542–544] Less often does one find a long contemporary narrative clearly describing a deep state of selfless "oneness," a state whose residues could then be said to continue in a form consistent with a sustained, major *Exceptional Stage of Ongoing Enlighted Traits*.[10]

The rare, extraordinary state of Being deletes, to absolute zero, all existential frames of reference. [ZB: 627–630; ZBR: 354–355, 392–393] It seems to reflect the complete dropping out of (1) all obvious prior personal and Self—other functions (these are discussed in chapter 5), and (2) all subtler functions that had been referable to the brain's widespread representations of place and direction (these are discussed in chapter 6).

Did Master Hakuin prescribe meditation in stillness on a cushion as the best way to become enlightened? The next chapter explores the living Zen alternatives.

3
Developing Traits of Character on the Way to Altruism

Meditation in the midst of action is infinitely superior to meditation in stillness.
Zen Master Hakuin Ekaku (1689–1769) (his final calligraphy)

Talent is formed in stillness, character in the world's torrent.
Johann von Goethe (1749–1832)

Meditation doesn't leave the mind totally blank. It supports a mindful introspection. This more objective instrument helps us develop character on the Path. [ZB: 125–129] During retreats, an unsparing introspection has ample time to diagnose our major imperfections. It also makes us keenly aware how much more hard work is required to dissolve them. This work occurs during living Zen, during daily life practice out in the world at large.

Yamada Koun-Roshi (1907–1989) once condensed the Path of Zen. To his student, Robert Aitken, he said: "The purpose of Zen practice is the perfection of character."[1] Aitken understood how much his teacher had just distilled into this one sentence. It was the six perfections of classical Buddhism (Skt: *paramitas*). [MS: 220 n7] In various translations, these are guidelines to a living Buddhist practice. They start with the virtue of generosity and continue as follows:

- Generosity of spirit in all matters
- Disciplined restraint of the passions (Skt: *shila*)
- Patience and tolerance
- Resolutely applied energy
- Meditative practices that dissolve the intrusive Self
- The ripening of authentic insight-wisdom

With regard to developing one's character in general, David Brooks's latest book is recommended reading.[2] *The Road to Character* explores the creative tension between two sets of virtues—those we might list on our resume, as contrasted with those more elevated virtues we hope that someone else will mention in our eulogy. Our resume-virtues are Self-centered, career-oriented, and status-seeking. They are dominated by the crass economic logic that drives a Self-promoting applicant. In contrast, our eulogy virtues are Self-effacing, oriented more toward altruistic service to others, and are often expressions of genuine humility. No "selfies" intrude in this category.

In our culture's dominant value system, what determines character? Mostly how we *wrestle* with our character flaws—at both conscious and subconscious levels—trying repeatedly to correct them (while trying also to maintain our latent virtues). Each prominent person in the West whom Brooks chose to examine "had to descend into the valley of humility to climb to the heights of character." ... "in the valley of humility, they learned to quiet the self. Only by quieting the self could they see the world clearly, ... understand other people and accept what they are offering."[3]

A key distinction becomes useful at this point. It differentiates overt conscious processing from *subconscious* processing. The conscious way to *correct* one's character is direct. It requires hard, disciplined *work*. This is an obvious, effortful, top-down job of Self-correction. In contrast, the subconscious way allows for the rough-and-tumble flow of a living Zen practice—out in Hakuin's "midst of action"—to wear down one's sharp-edged projections. This is a subtler process of retraining. This approach allows room for one's behavior to evolve gradually, by *indirection*, during the give-and-take friction of daily relationships with other persons. These everyday lessons help us adapt our imperfections at less conspicuous subterranean levels and with less effort on our part. [MS: 130–150]

Two of the subjects in Brooks's books are Dwight D. Eisenhower and George C. Marshall. Although both men had trained in a rigorous military tradition, they began with very different basic personalities. For Eisenhower, the fully conscious task was to conquer himself. For Marshall, the fully conscious task was to master himself. Each general, often struggling consciously to correct his individual flaws, would become capable of inspiring countless others to monumental heights of sustained performance during wartime.

Cultural Estimates of Character, East and West

In this twenty-first century, we in the West have different value standards for the merits of reticence versus those of all-out self-exposure. We weigh our judgments of character and wisdom on different scales than those that had guided Zen Buddhists during earlier centuries. [ZB: 641–645] Two well-known literary men grew up in the midst of the military atmosphere in the Japan of their era. Yet each one was still admired for the restrained, natural simplicity of his lifestyle and for his extraordinary poetic gifts.

For example, Ryokan (1758–1831) was a venerable Soto Zen priest, a wandering mendicant sage, and a skilled calligrapher. He exemplified "the serenity of nature itself, in his boundless love and thoroughgoing compassion. The nobility and richness of his character were the outcome of long, strict Zen practice that had cleared away the vines and creepers encircling his mind. This is the secret of the lasting affection in which Ryokan has been held by all."[4] As for the reasons the second man is held in lasting affection, much more is said about Basho in part V.

What Can Zen Buddhism Offer Today?

In the West today, Zen offers less of a military boot-camp atmosphere than it did in centuries past. It remains a firm, disciplined presence, but one now relatively gentle in the uphill task of dismantling the pejorative Self. [ZB: 141–145] Its trainees are also more openly responsive to the kinds of ethical governance they can resonate with from their leaders, both men and women.

Aitken-Roshi illustrates how these subtle mutual influences operate in a *sangha* by citing what he heard from the lips of Shunryu Suzuki-Roshi.[5] Suzuki said this about his own sangha members: "I don't do anything. I just sit there and the members organize themselves. It is not my doing." Yet his leadership style kept inspiring other members to do the hard work of buying the buildings, establishing first a zendo, then a mountain retreat, and so forth. Good things kept happening.

Aitken explains how such influences happen by *indirection*. In one sense, they develop "by virtue of one's virtues." Why do they tend to just "happen"? It is because they "mirror" the spontaneous affirmative qualities of character that an excellent role model is expressing naturally, *subconsciously*. [SI: 76–78] Despite the sentimental overtones that the word *virtue* can have, "It is indeed virtue that we try to uncover in our practice."[6] He describes what develops when one practices in a way that starts to unveil the essential core of reality (the way all things *really* are). Now what happens is "teaching without overtly teaching. You forget yourself and trust others, in effect, revealing to them their own essence."[7] This subtle aspect of character resists being localized in an MRI scan and eludes the printed page.

It is in this sense that Aitken realizes what Yamada Koun-Roshi was *not* saying. He was *not* saying that *we* (our *I-Me-Mine*-Selves) are the ones who consciously "correct" *our own* character.[8] Instead, what his teacher was really pointing to was that further refinement of advanced Zen practice. At this later stage of retraining, one's "body and mind fall away, and the sparkling-gemstone of character stands forth pristine."

Is all this one more example of spiritual overbelief? Of students' overworshiping their spiritual teacher? I don't think so. Both as a teacher and a student of neurology, I believe we've now come to the presence of a core quality of leadership. It is a quality that helps define an inspiring Buddhist master, a general, a parent, or any other exceptionally skillful teacher. Beyond all soft notions of surface charisma, they excel as *educators*. [ZB: 119–125] Dictionaries remind us that the root origins of the verb *to educate* mean *to lead forth* each *individual student's innate essential nature*. How do the best teachers educate others? By setting inspiring examples of their own genuine enthusiasm, virtuous character, and competence. Like prolific fruit trees, their influences can be understood in terms of those fruits to be borne in subsequent generations. Yet ultimately, whether these factors become effective in any era depends on how an educator, or a muse, motivates each individual student's native capacities.

Native Capacities

Daisetz T. Suzuki brought Zen to the West. He knew that authentic Zen teachings do not try to destroy the warm prosocial factors in the affective core of the human personality. Instead, Suzuki emphasized that the teachings aim to eliminate our errors of discrimination and unjustified assertions. When these drop off, and our mind finally becomes clear, then, he said, "the heart knows by itself how to work out its native virtues."[9]

These *native* virtues then reflect the original pristine resources of perfection that took millions of years to evolve in the depths of the human brain. These are the sources of our so-called original "Buddha nature." In the small brain of a nesting bird, comparable neural networks drive migratory flights for thousands of miles, to the South or North, at just the correct time every year. Like the root origins of genuine human compassion (Skt. *karuna*) the latent resources of wisdom coded in such instinctual avian networks could only be refined, step by hard-earned step, during the course of evolution. [ZB: 683–691] In today's aberrant culture, our native human capacities may also take long decades to be accessed, retrained and expressed more fruitfully.

Do the Zen Buddhist teachings rest on a sound ethical foundation? [SI: 202–207] In my case, one of *kensho's* realizations arrived transparently clear in 1982. This insight took the form of a Utopian message: *If a person could maintain this same selfless state, such a perspective could serve as the basis for all ethical behavior.* [ZB: 539; ZBR: 398–401] That Zen Buddhist practice did have such a deeply ethical core was already apparent at least 14 centuries earlier. Then the sixth Zen patriarch, Hui-Neng (638–713) characterized the selfless, errorless rectitude of our original Buddha nature as being "without error, disturbance, and ignorance."

Altruism

When Kobori-Roshi first introduced me to Zen, he pointed to the fact that selfless insight-wisdom "does not exist for itself. What it leads to ultimately is to sharing, to compassion, to love." [ZB: 64, 648–653] Altruism is this compassion in action. It is our generous, selfless giving for the benefit of *other* beings (Old French *autre* = other). [ZB: 691–695]

The most appropriate contemporary spokesperson for a major new book on altruism is Matthieu Ricard. [ZBR: 397–398; ZBH: 170–172] His subtitle is accurate: *The Power of Compassion to Change Yourself and the World.*[10] Those who may question the existence of genuinely selfless behavior will find examples of spontaneous altruistic behavior among animals (pp. 183–207) and children (pp. 208–224). Those who worry realistically about the fate of our planet can be encouraged by the example set by this remarkable Tibetan Buddhist monk and by the facts documenting Ricard's global perspective.

The 19 altruists studied by Marsh et al.[11] formed a distinct subset of empathic and compassionate individuals. [ZB: 650–653, 693–694] They were selected because they had donated one of their own kidneys to a stranger. Moreover, these altruists also showed enhanced abilities to recognize the expressions of fear on other persons' faces. Their amygdala was not only larger on the right side. It also showed enhanced fMRI signals when the subjects responded to other persons' fearful expressions. This set of findings placed the [surgically defined] altruists off at the "high end" of a multidimensional, pro-social continuum. Notably, at the other extreme were the data and traits of uncaring psychopaths.

Recent Interviews with Contemporary Buddhist Teachers in the West

Richard Boyle is a sociologist and long-term Zen practitioner. He has just published the instructive results of his interviews with eight men and three women who are currently engaged in an active role as Buddhist teachers.[12]

Boyle summarizes the three main features of these teachers' own meditative practices as:

- Controlling attention and quieting the mind
- Letting go of prior conditioning
- Cultivating compassion

He then summarizes the phenomena that characterized their acute state of awakened consciousness. Their descriptions fall into three representative clusters.*

- This awakened state did *not* add something. Instead, it deleted a key aspect of their ordinary perception—the pervasive influence of an intrusive Self.

---

* Boyle's discussion chapters further illustrate why *states* of awakening are different from that ongoing *stage* that can be called the Exceptional Stage of Ongoing Enlightened Traits. The teachers themselves do not consider themselves 100 percent perfected human beings (nor are the other recent teachers discussed by Boyle). Awakening does not guarantee genuine lasting compassion or sainthood.[13]

- This state was devoid of emotional attachments to the Self.

- Awareness and prompt action were coarising spontaneously.*

The take-home message is that "awakening" for these teachers is not a final destination. Enlightenment is a work in progress, as we saw in chapter 2.

As we now move into part II, the next chapters clarify why one's emergent positive traits of character can develop fully only when those old, inflexible boundary relationships that had separated Self from other are transformed.

---

* Behavioral spontaneity can be released in different ways during internal absorption and *kensho-satori*. [ZB: 668–677]

Part II

Implications of a Self–
Other Continuum

In order to discover who you are,
First learn who everybody else is.
You're what's left.

Fortune cookie message, 2010

The Self: A Primer

> To study the way of the Buddha is to study your own self.
> To study your own self is to forget yourself.
> To forget yourself is to have the objective world prevail in you.
> Master Dogen Kigen (1200–1253)

Before we examine what a living Zen is in the next four parts, we first need to understand *who* engages in it. For many centuries Zen has emphasized two major themes: the selective targeting of our overconditioned Self and the training of attention.[1] As this new millennium dawned, functional imaging results showed that *the relationships between the Self and attention are often reciprocal.* [SI: 98–108] One central fact is important for all who meditate: our brain's natural capacity is to express these relationships *inversely*—as one increases, the other decreases.

The Semantics of Self

To Freud, the *ego* was our pragmatic executor. We depended on our ego's many positive attributes, including its introspection and logical reasoning, to help channel the passionate instincts of our *id* along socially appropriate lines. [ZB: 34–36] Unfortunately, in common speech, this word, *ego*, later acquired very different meanings. For example, the person who was too selfish was often said to have an "inflated ego." Please notice: meditative training is designed to liberate the *original* Freudian ego and to enhance the adaptive competence of its pragmatic Self. What the training process will diminish, at the same time, is that old, overinflated, *maladaptive* egocentric Self. Notice the paradox. Our Self does help us to be defined and to act. However, it is essential to know when and how to let go of it. So let us review the nature and extensions of this construct of Self because themes related to Selfhood will figure prominently in this and the next three parts.

Meanwhile, we each keep polishing the halo around this personal Self, grossly underestimating how pervasive are its underlying Self-centered liabilities. Simply as a way to emphasize the hidden numbers, size, and strength of our liabilities, these pages continue to spell the Self with a capital S.

A *somatic* Self does exist in the here and now. Our own bodies provide tangible, proof-positive evidence. [ZBH: 12–22] Moreover, our own thoughts provide evidence in support of a *psychic* sense of Self. Our psyche is almost entirely intangible. Although we can't touch any thought, our body can register the feelings associated with becoming fearful, angry, or sad.

Our Self seems to exist as we travel through time and across distances. Its sense of agency lingers within each ongoing second of our awareness and in our short-term working memories. We establish our own personal history in layers of long-term memories that extend private episodes far back into our childhood. We embed ourselves in the notions of plans that we project into the distant future.

Looking further into our implicit psychic sense of Self, we can identify the outlines of its three functional components. [ZB: 43–47] Their corresponding operative words are *I*, *Me*, and *Mine*. Each component is burdened by maladaptive aspects that cause many problems.

- The *I*. This *I acts* in a sovereign manner. Other persons are keenly aware of its transgressions.

- The *Me*. This *Me* is vulnerable. It reacts in fear and anger, because other things and other people can bruise it, physically and mentally.

- The *Mine*. This *Mine* possesses its own turf. It grasps and holds on to external material possessions. Internally, it clings to rigid opinions and clutches fixed habit patterns.

Where Is the Self?

One might begin to think of the Self as huge constellations of linked modules distributed in networks throughout the brain. Operating together, they become a dynamic, omnipresent, somatic and psychic construct. [ZB: 40–42; ZBR: 11–19; SI: 53–83] The following narrative paragraphs summarize the neural substrates of this large new field. Viewed charitably, it remains a work in progress.

Neuroimaging made impressive advances during this first decade and a half of the new millennium. [ZBR: 187–191; SI: 251–268; ZBH: 154–170] Several regions emerged as candidates for our normal autobiographical, Self-referential psychic functions. [ZBR: 193–200, 204–207] The two largest regions were on the *inside* of the frontoparietal cortex.

Likewise, the functional anatomical correlates of attention were also being defined. Studies pointed to a *dorsal* and a *ventral* system of cortical attention. Each system was distributed over the *outside* frontoparietal surface of the hemispheres. Using two old Greek words, our attentional processing could now be divided into two categories: (1) *Ego*centric attention. Its targets were referred back to the axis of our somatic and psychic Self; and (2) *Allo*centric attention. Allo- simply meant *other* (In ancient Greek, *állos* = other). This form of other-referential attention was being directed toward items *out there*, in *extra*personal space. Please notice: this allocentric attention operates *anonymously*. Its lines of sight are not immediately referred back to us in some Self-centered location.

Now the stage was set for functional magnetic resonance imaging (fMRI) to monitor the ways brain signals *fluctuate* when subjects were scanned in their so-called "resting state." This state was first considered to be relatively neutral and "task free." In reality, as meditators know only too well, it was mostly preoccupied with Self-referential mind wandering.

Again, as the earlier PET scans had shown, much of the brain's "resting" fMRI activity was found to originate in the so-called "default system." [SI: 70–76] Hesitantly, its many Self-referential functions became increasingly appreciated.[2] Yes, those two largest, most metabolically active regions of this system were represented deep along the *inner* surface of the brain. There—front and back—they occupied (1) the medial prefrontal cortex (MPFC) and (2) the medial posterior parietal cortex. This medial posterior region was the larger. It included the *posterior cingulate cortex*, the *retrosplenial cortex*, and the *precuneus*. [SI: 61] A third constituent of the so-called default network was the *angular gyrus*. This small gyrus was located on the brain's *outer* surface. It occupied the posterior part of the inferior parietal lobule (chapter 15).

Most striking now were other findings. Under conditions that were either reactive or spontaneous, this usually active default system could become *deactivated*. Indeed, these two similar *de*activations occurred whether (1) a fresh task or stimulus immediately captured the person's attention, or (2) an activation occurred *spontaneously* in the brain's normal frontoparietal attention-processing systems. [SI: 98–108] Notice that this second, spontaneous, reciprocal cycle recurs *slowly*, only three to four times a minute. The neural-metabolic mechanisms that generate this slow reciprocal cycle have not yet been determined.

These two observations have substantial implications. The suggestion is that deeper subcortical mechanisms must exist. The deep, central position of the thalamus renders it a plausible candidate for the capacities that could shift two such major functional categories—Self/attention—in inverse, reciprocal directions. [ZB: 263–274] What else makes our recognition of these two see-saw phenomena so important? These reciprocal shifts can help clarify how some triggering stimuli and deeper intrinsic mechanisms can combine to precipitate *kensho-satori* and related states of selfless awakening. [ZB: 615–617; SI: 109–117]

Recent Studies of Our Normal Autobiographical Self: A Progress Report

During the past decade and a half, Hans Lou and colleagues in Denmark have issued a series of reports from Aarhus University.[3] [ZBR: 204–206] In brief, the weight of this evidence supports the proposal that the core of our reflective Self-awareness hinges on a dynamic, two-way continuum of connections. These link the medial frontal, medial posterior parietal, and angular gyrus regions *with the thalamus*. Clearly, other functions are also attributable to the neurons in each of these heterogeneous regions. Notwithstanding, our hippocampal nerve cells also appear normally to engage with networks of posterior parietal and other medial temporal nerve cells in configurations that express a distinctive blend of "Self-othering" functions[4] (chapter 6).

Studies Suggesting That a Diminution of Self-Referential Activity Is Related to Certain Later Phases of Meditation

The early PET studies from Denmark in 1999 showed that prefrontal and cingulate activity was substantially *reduced* when the subjects let go of their *willful self-control* during the *later* phases of meditation.[5] [ZBR: 216–218] Today's neuroimaging studies of meditators are meeting increasingly rigorous standards in ways that confirm and extend these important pioneering findings.[6] [ZBR: 214–226]

Why was so much of the earlier fMRI and behavioral research on *"selflessness"* difficult to interpret? It was because the experimental design included an inherently intentional, *task-based voluntary* structure. This was necessarily imposed from the top down. [SI: 39–43] We need to remind ourselves of a crucial fact: *we impose* some kind and degree of *our own* attention whenever we perform a task, especially a complicated cognitive task. Our attention is the sharp point out at the tip of a long arrow. The shaft behind it represents the bulk of our neural processing. Attention remains the vanguard that impales the target and *leads* our processing. [SI: 14–34, 98–103; ZBH: 22–33, 220–221]

A recent MEG report becomes germane[7] (see also appendix C). This study of experienced meditators highlights how diminutions in the Self are accompanied by *reductions* in their *beta* and *gamma* frequencies. These 12 long-term Vipassana meditators had an average of 16.5 years of practice. Two of their phases of meditation are relevant. Let's first consider their later, deeper phase. They called it "narrative self-awareness." The *diminution* of this Self-referent phase correlated with a *decrease* in high-gamma (60–80 cps) activity, especially in the *left* hemisphere. The gamma decrease was evident especially in the *left* frontal, thalamic, and medial prefrontal regions, both ventrally and dorsally. These findings suggested that the meditators' language networks—those *left-sided* networks normally responsible for their discursive word-thoughts—had become progressively *less* active during the *deeper* phase of meditation. [SI: 150–152; MS: 145–150; ZBH: 132–133]

Next, let's examine their earlier phase of "minimal self-awareness." The *diminution* of this earlier meditative phase was correlated with *decreases* in *beta* band activity (13–25 cps). The distributions of these early beta decreases overlapped those of the later gamma decreases cited above, in both the left ventral medial prefrontal cortex *and* the thalamus. Moreover, beta decreases also extended into [such Self—other autobiographical regions as] the right posterior cingulate cortex, the right precuneus, and the inferior parietal cortex (R > L). Notably, the four most expert meditators showed the greatest beta reductions in both their right inferior parietal lobule and left dorsal medial *thalamus*.

These references to the thalamus in this recent MEG study serve as a signal. They remind us that deep covert thalamic mechanisms are pivotal to our understanding of normal and meditative states of consciousness. [ZBR: 167–179; SI: 87–103] These are considered later in chapter 6. But first, as you arouse from sleep each morning we need to examine why you, the reader, know *that* you are, and *where* you are, in the space outside your body.

5

Emerging Concepts in Self–Other Relationships

> The ventral pathway is an occipitotemporal network that bridges the early visual cortex and specialized cortical and sub-cortical structures involved in various forms of memory and learning—specifically habit formation, emotion, long- and short-term memory, reward and value.
>
> D. Kravitz and colleagues[1]

Each of these five functions is already a big topic. This chapter adds a sixth: Self/*other* frames of spatial reference. Can one ventral pathway, plus its extensions, support all of these vital computations? Be prepared for evidence that comes from monkeys, normal human subjects, and patients who have neurological lesions.

The way *kensho* struck on the London train platform back in 1982 made one thing clear: *kensho's* mechanisms needed to be clarified in ways that explained how they could cut off the Self.[2] [ZB: 536–539] In the neuroscience literature, a remarkable notion was slowly emerging. It served to remind us that Immanuel Kant (1724–1804) had once proposed a similar concept: there were *two* ways of looking at the whole world—*our* own personal way, and a more independent [anonymous] way. [ZBR: 361–363]

The previous chapter prepared us for some Self/other distinctions. Yes, the obvious way was our familiar dominant, Self-centered way. The other way required us to shift into a whole new mental set. This second way not only drew on the alien Greek word *allo*, it also pointed into the reality of the external world—*out there*. Strangest of all was its stark anonymity. This other way of pointing outward to an item in the environment had no *initial* need to refer its percepts back through the lens of some subjective viewer who always remained centered in the axis of Self. [ZB: 491, 523, 801; ZBR: 370–371; SI: 118–120; MS: 171–172; ZBH: 26–33][3]

Of course, there were always those odd witnesses who alleged that their sense of Self had dropped off during *kensho*. Yet researchers in general seemed unaware that a connection could exist between this direct ego- to allo- shift (seemingly in the marrow of one's bones) and the deep awakenings that had arisen at the historical core of the Buddhist Path. [ZB: 549–553]

In fact, studies of single temporal lobe nerve cells in primates had already revealed intriguing results as far back as 1972.[4] Why were Gross and his colleagues so surprised by their findings? Because these nerve cells were very talented and sophisticated. For example,

- The nerve cells in their monkeys' inferotemporal cortex showed remarkably large visual receptive fields. Their mean size was 409 degrees squared. Seven of the eight cells shown in figure 2 of their report had response fields that covered areas on *both* sides of the vertical meridian. The largest of these areas was approximately 685 degrees squared. About 47 percent of this one neuron's already huge left field had extended past the midline, enabling it to also respond to stimuli over in the *right* (ipsilateral) field! Clearly, these neurons in one temporal lobe were tuned to receive (and coopt) some visual input that had crossed over from nerve cells in the opposite hemisphere.

- "The center of gaze, or fovea, fell within or on the border of the receptive field of every inferotemporal neuron studied" (chapter 6).

- Meaningful biological stimuli proved most effective for these nerve cells. These stimuli were profiles of human hands or monkey hands that showed the fingers. Simple circles or bars proved less effective.

Finally, in 1990, Tamura and colleagues[5] presented both visual *and* auditory stimuli to monkeys who were fully awake. They found that some neurons in the hippocampus responded differently to stimuli in egocentric space than in allocentric space. For example, by rotating the chair in which the monkey was seated, it was possible to test the effects of environmental cues on its brain's direction responses. Now, 6 of these 17 neurons responded to directions that remained constant *relative to that monkey*. Accordingly, they considered these neurons to be egocentric (*self*-referential). The directional responsiveness of four other neurons remained constant *relative to the environment*, independent of that monkey's position. They regarded these as allocentric (*other*-referential) neurons. Clearly, these allocentric responses were coded for one fixed location *out there* in the environment. This was not a direction established by its relation to that monkey's head or body.

Later experiments recorded from awake monkeys who were now free to search for objects in a complex natural scene. Now, the receptive fields of their inferior temporal cortex nerve cells were much smaller.[6, 7] Moreover, certain nerve cells seemed to be at a particular stage in their visual processing that was beyond the reach of emotion. Indeed, these neurons responded to visual stimuli "independently of their reward or punishment association."[8]

In short, these neurons seemed capable of an almost *objective* form of vision. Their perceptions were free to proceed without being contaminated by subjective, emotional ties. Why might reading these words in this article "ring a bell" in me (so to speak)? It was because they reminded me of a pivotal event that happened more than three decades earlier. Then, "objective vision" was the phrase that rose spontaneously to describe the emotionally *un*attached visual perception that had just struck off the Self during *kensho* [ZB: 538, 573–574] (appendix G).

In order to pursue such differences, researchers knew that they needed to mimic some elementary qualities of the natural environment if they were going to test which basic skills the brain used to perceive what is "out there." Accordingly, Vaziri and colleagues[9] presented monkeys with two categories of stimuli. Some stimuli were totally enclosed like a ball (resembling a real fruit), not open like a bowl. Sixty-five neurons in the ventral part of the superior temporal sulcus (STS) did respond chiefly to items that had this *enclosed* shape. Notably, a major projection runs forward from this same temporal lobe sulcus (chapter 15). This tract carries messages to regions that infuse some sense of *value* on the item that is being visualized. (Monkeys recognize some fruits as sweet, others as sour.) These value-conferring frontal regions include the ventral lateral prefrontal cortex (IFG) and the orbitofrontal cortex.

For contrast, 76 other neurons responded to stimuli that represented the *unbounded* 3-D space of an *environmental setting*. Mimicking natural landscapes, this scenery was so wide open that it extended beyond the monkey's field of view. These cells occupied the dorsal part of the anterior inferotemporal gyrus. The findings become of interest, in light of the huge field of ambient vision that can open up space when one drops into a state of internal absorption. [ZB: 495–499]

Emerging Notions about the "Self" in Normal Human Subjects

Back in the early 1960s, suppose some committee had been charged with describing the general characteristics of the "altered" states of consciousness. It might still have managed to get by without having to grapple with a big concept such as the "Self." [ZB: 309–311] However, by 1975, the Self seemed to have "grown up" in stature. Then, a committee *was* set up expressly to describe such states of consciousness. After deliberating for a whole morning, the members finally realized that they needed to define specific properties related to the concept of Self. Among the reasons for this was their recognition of a glaring fact: Selflessness could occur. A person's entire sense of Self-control over events could actually drop out at certain times.

As a useful concept, the Greek word *allo* has also been slowly appreciated within the field of neuropsychology. Yet its "other" frame of spatial reference still remains counterintuitive to every person's dominant egocentric view of the world. Just recently, in 2012, Chin required 18 pages in order to conduct a tutorial review of object-based attention[10] (appendix G). In 2015, Foley et al.[11] used eight pages to review a theory of Goodale and Milner. In 1992 they had proposed two visual frames of reference to explain the different ways that we *attend* to reality. One way was called "a vision for perception." The other way was "a vision for action." Other researchers have been using at least eight different word pairs to help differentiate other-referential (allocentric) attentive processing from Self-referential (egocentric) attentive processing. [SI: 53–64]

Could some of this difficulty reflect the fact that single neurons, each representing one or the other of these two different frames of reference, are sometimes actually closer together—anatomically and in their physiological linkages—than the old Greek words might suggest?

The Two Large Medial Cortical Regions

Earlier clinical deficits and basic research had pointed long before toward the *retrosplenial* region, back in the medial *posterior* parietal cortex. Its exact boundaries and functions seemed elusive. [ZBR: 73–74, 85] One study in 2012[12] followed 13 subjects who underwent rigorous tests of their navigation abilities in a virtual maze. The term, "perceived heading direction" was used. This phrase, like the term "head direction nerve cells," could be confusing in the context of this experiment.

Using a joystick, the subjects started from the center of the virtual maze. They then navigated toward various landmarks positioned out at its periphery. The direction they *first* faced became thereafter the *equivalent* of their ongoing, *subjective* "North." This direction automatically established that those remaining three directions would be subjectively perceived as 90 degrees apart. Only one brain region—the *left precuneus* (BA 31)—correlated with learned changes in these heading directions. In fact, this cluster lay posterior and dorsal to the actual preferred anatomical location of the retrosplenial cortex. This standard location for the retrosplenial cortex is defined as BA 29 and BA 30. It lies immediately behind the splenium at the back of the corpus callosum. [SI: 25]

Additional research reports gradually made clear that:

- Many *in*tangible Self-relational functions referable to an autobiographical psyche were being represented more anteriorly, namely up forward in the *medial* prefrontal cortex (MPFC).

- Here, the Self-referential *internalized* judgments tended to activate the *ventral* portion (VMPFC). In contrast, judgments about close personal friends and relatives more frequently activated the *dorsal* portion (DMPFC).[13] [ZBR: 193–200]

- This medial prefrontal cortex usually *coactivated* seamlessly with its posterior mostly *medial* parietal counterparts.

- Much of this large medial parietal region, in turn, kept registering (and appeared to help retrieve) information chiefly representing the local topography and *related* circumstantial details that came from the *other* world outside the Self.

- The operational result often appeared to be a *joint* construct, a consortium, a Self-in-the-world continuum.

In this joint autobiographical role, the brain appeared to keep an ongoing record of separable events that took the form of almost journal-like personal entries. These entries became episodic memories. For example, each episode established the memory that *My* private Self had occupied one particular place at a certain time (chapter 6). How could a brain accomplish so seamless a merger of its medial frontal and posterior parietal networks? Only by blending them into a spectrum of *"Self-othering"* functions. [SI: 58–59, 72–74]

Years ago, a spontaneous "quickening" released some of these same joint memory functions into conscious images while I was *meditating* one evening. The result was an unrelated series of rapid tachistoscopic images. These represented brief, random episodes retrieved from previous decades. [ZB: 390–391] A yoga teacher recently informed me that she had experienced similar phenomena during a Vipassana retreat.

A Model of Medial Parietal-Temporal Processing

Recently, researchers challenged the spatial working memory of 15 male "gamers." Their spatial memory task was to recall *where* different objects had previously been located in a virtual scene.[14] The subjects' point of view could be rotated, or otherwise changed, in terms of its relative direction. To explain the resulting data, the authors presented a complex working hypothesis. The model serves our purpose here as a way to introduce the question that is discussed further in later chapters: How do all these functions far back in the posterior *medial parietal* cortex join together with our *medial temporal* lobe functions to access and process these memory-based remindful tasks?

Their model postulates that:

- The (*outer* surface of the) posterior parietal cortex helps both to form an egocentric representation of external landmarks and to establish where their boundaries are.

- The *medial* posterior cingulate/retrosplenial region of cortex helps to develop a "map" that represents landmark locations. These constructs can be further modulated by *either* egocentric *or* allocentric frames of reference.

- When are these constructs modulated by egocentric functions? Especially when, at *that* very moment, the circuit is being driven by actual, fresh incoming *sensory* information.

- When are these constructs modulated by allocentric "head direction" functions? When the circuit is being driven by the *retrieval* of a relevant memory from the *past* (appendix C).

- The parahippocampal cortex also contributes to the formation of an allocentric map of landmark locations (chapter 15).

- How do cells in the hippocampus respond to different place locations? By encoding "*conjunctions*" of their landmarks, boundaries, and other contextual information (chapter 6). The dictionary definition of a conjunction implies that things are being joined together in a unified association (appendix G).

The items in the above formulation are instructive. They are compatible with the evidence that some allocentric and some egocentric functions co-exist in adjacent or separate parts of the brain (chapter 6). However, note that these separate modules can still be connected, integrated, and "conjoined" by myelinated fiber (white matter) pathways. This perspective keeps open the prospect that pathological lesions that damage different locations—including those that disconnect white matter tracts—can cause neurological deficits. These clinical deficits would then represent *dys*functions that are either egocentric, or allocentric, *or* combinations of the two.

Cortical Functions Referable to Regions over the Outside of the Brain

Our normal constructs of Self extend well beyond the reach of lesions restricted to the *medial* prefrontal and *medial* posterior parietal cortex. [SI: 135–141] For example, a recent study of 15 normal subjects[15] serves to introduce here the reasons for our later focus on the *angular gyrus* (chapter 15). This is a relatively small gyrus (BA 39). It lies on the *outside* of the brain, in the posterior part of the inferior parietal lobule. On the right side one of its roles is to convey—in *metric* terms—several vital constructs involving distances. One is our sense of "relative egocentric distance." Another is our sense of spatial, time-related, and social distances. Its degrees of activation can also be correlated with such metric distinctions as those between "close friends/distant relatives" and "near future/distant past."

The right angular gyrus lies close to the *right* temporoparietal junction (TPJ). This large TPJ region is a part of the *ventral* frontoparietal attention control network. In the conventional role currently assigned to it, one potential function of the TPJ is to help the brain "switch" its attention. These reflexive switches can shift attention back and forth (1) from processing earlier, low-level, more concrete representations, (2) to processing more abstract and contextualized representations that are then being grounded in higher-level constructs of our Self-identity. [MS: 17–18, 25, 105, 107, 161; ZBH: 138, 205 n6, 225–226 n1]

The whole frontoparietal control network is extensively distributed over the outside surface of the brain. It services our personal needs for attentive processing, chiefly in a more Self-centered manner. The large middle cerebral artery is our one major source of blood supply to much of the outside of each hemisphere. Unfortunately for attempts to localize discrete cortical functions, branches from this large artery supply blood not only to the cortex, but also to the underlying white matter, and to subcortical regions. This poses big problems. It means that when an embolus or thrombus occludes this middle cerebral artery, it can underperfuse, infarct, and/or disconnect several very different brain regions. Moreover, current neuroimaging techniques still commit errors in localizing stroke lesions.[16] Mislocalizations are often related to the normal variations in the blood supply represented by different branchings of the underlying arterial tree.

For these several reasons, neurologists have been having a difficult time defining precisely what ego- and allo- mean in terms of each individual patient. The following narrative sections briefly summarize the current status of what is still another work in progress.

Hillis and colleagues reported a classic study in 2005.[17] Twelve patients had an egocentric visual neglect only; four patients had an allocentric visual neglect only. Each had a relatively small, *partial, acute* ischemic *cortical* lesion. This was documented by sensitive *diffusion-weighted imaging* (DWI) (appendix D). Please notice that these were *not* regions of actual *infarcted* cortex. Instead, DWI defined what were only discrete areas of *cortical hypoperfusion*. The four instances of allocentric neglect correlated with this underperfusion of the *right* superior temporal gyrus (BA 22). In contrast, most instances of egocentric neglect were strongly associated with *parietal* hypoperfusion. This involved the right angular gyrus (BA 39) and right supramarginal gyrus (BA 40). Additional egocentric neglect correlations were associated with hypoperfusion of the right posterior inferior frontal gyrus (BA 44) and the right visual association cortex (BA 19).

This 2005 study had another advantage. It included the Ota Test. Why did this test help confirm that the patients allocentric deficit was a so-called *object* type of neglect [SI: 64–70]? Because these patients neglected that *lateral* half of any discrete *object — whether it was out in external left space or right space —* the *lateral* half that was out there on the opposite side of the brain lesion (appendix G). In contrast, a patient with an egocentric neglect of attention neglected only the items in each half of the *subject's* visual fields that were on the side opposite that *subject's* brain lesion.

An earlier study in 2003 had already reported that five patients with right *posterior* cerebral artery infarcts had shown neglect.[18] At that time, all of these patients were thought to have this visual neglect only in their opposite visual field. No Ota test had been performed. The infarctions in these right posterior cerebral artery distributions all extended far enough forward to include the medial temporal lobe in its inferior portion. Here, the infarctions were centered especially on the *parahippocampal gyrus* (chapter 15). Indeed, the right parahippocampus was the critical region for left visual neglect in this report. The six other patients with posterior cerebral artery infarction whose parahippocampus was *spared* had *no* visual neglect. An infarct in the *right angular gyrus* was the most consistent finding in 14 of the 24 patients whose middle cerebral artery was occluded. The superior temporal gyrus was also involved in seven of the neglect patients.

Given these right superior temporal lobe lesions, plus the ventral location of the right parahippocampal lesions, it remains possible that several allocentric field defects could easily have escaped being diagnosed in this earlier study of neglect.

In the 10 cases of posterior cerebral artery infarction then reported in 2006 by Bird et al.,[19] the presence of visual neglect was correlated (using diffusion-weighted imaging) with damage to one white matter tract in particular. This tract (the *inferior longitudinal fascicle*) connects the parahippocampal gyrus with the angular gyrus higher up on that same side. Once again, the ego- or allo- nature of the neglect was not specified. However, two of the seven patients with neglect also had damage to the parahippocampal gyrus in the opposite hemisphere.

By 2008, it seemed that most evidence would support a gross distinction between one group of patients who had a more dorsal lesion and another group who had only a more ventral lesion.[20] In that year, for example, it was stated that:

- "Disturbances of ventral (temporal) information processing—concerning detailed object representation—lead to allocentric impairment."
- "Disorders of the fronto(parietal) processing stream—dealing with spatial information—cause egocentric deficits."

However, four years later in a large study from Beijing by Yue et al., 30 of the 47 subjects with right-sided brain damage were considered to show *both* allocentric *and* egocentric neglect.[21] Ota-type tests were performed. In these neglect patients, most of the left allocentric neglect appeared to correlate chiefly with damage to the right superior temporal gyrus, the middle temporal gyrus, *and* the globus pallidus and putamen on that right side. These two nuclei of the basal ganglia contribute to the reflex pathways that begin down in the superior colliculus (chapter 14). Lesions here complicate the interpretation of this report.

In their comprehensive review of this complex subject in 2013, Humphreys et al. usefully summarized "object-centered" space. They defined it as being coded in the frame of reference of the main axis of the *object*, not of the *subject*.[22] However, they reached a sobering conclusion: there now existed a "lack of consensus" about whether (1) all forms of spatial representation were ultimately related back to the body or (2) there also existed other frames of reference abstracted from the body that represented spatial codes called allocentric, object centered, or environmental.

Suppose we were now to ask: Could some people normally be more Self-centered, whereas others are more other-centered? In 2015, Goeke et al. studied 1823 subjects from different countries.[23] North Americans preferentially used an allocentric strategy. Latin-Americans preferred an ego-centric navigation strategy. It could be interesting to see if this individual physiological bias varied over time. For example, if the tests were applied to a meditating population and to an appropriate nonmeditating control population, they might document a potential shift from an ego bias toward a more allo bias, both before and after *kensho-satori*. [ZB: 82–83]

Recent Studies of Normal Subjects

In 2014, Torok et al. conducted a behavioral study[24] on 50 normal subjects. Their task was to navigate in a bounded virtual reality computer environment without using landmarks. Subjects who took a ground-based perspective tended to perform using their egocentric frame of reference. Those who took an aerial perspective tended to employ an allocentric frame of reference.

Saj et al. monitored 16 normal subjects with functional MRI.[25] Their task was to evaluate the way a bar was aligned *either* in relation to the middle of *their own body* (egocentric) *or* in relation to another stimulus out there in the external environment (allocentric). [SI: 58–64] The authors concluded that, during the allocentric task, the subjects used a "viewer independent coding of spatial relations." This coding—*between* objects—was being mediated by the *ventral* attentive processing stream, and perhaps more specifically by the *left inferior temporal gyrus* (chapter 6). In contrast, egocentric tasks activated a much wider area. These tasks emphasized the role played by activations throughout the *right* hemisphere in its occipital, parietal, and inferior frontal regions as well as in the precuneus and supplementary motor area.

In 2015, Lin et al.[26] used an independent component analysis (ICA) method to reconstruct EEG data on 21 young Taiwanese men (appendix E). These EEG-monitored subjects then performed a path integration task in a virtual 2-D open maze-like environment. During allocentrically oriented navigation, their 8–13 cps alpha desynchronized activity was followed (during and after turning) by bursts of synchronized 12–14 cps EEG activity (chapter 12). Both changes were referable to a region they called the "retrosplenial complex" (BA: 29, 30, 23, 31). Egocentric navigators showed stronger theta power increases in their medial frontal cortex and beta increases in their motor cortex. Allocentric navigators showed stronger alpha modulations posteriorly, in the retrosplenial, parietal, and occipital cortex.

In their meta-analysis of fMRI studies of normal human navigational skills, Boccia et al.[27] introduced a further issue. What influenced their results was how long each person's memory had been stored. Older memories that dated from long-familiar environments tended to be processed by an extended temporal-frontal network. In contrast, their shorter-term memories were based on recently learned environments. These tended to activate the parahippocampal cortex and the parietal and occipital lobes.

The important review by Ekstrom and colleagues[28] draws the following conclusions: Even "when participants make judgments regarding relative distances or directions of objects to each other, egocentric representations still serve as important anchors in cues in solving these tasks." Humans activate multiple brain regions when they use allocentric computations, "including parahippocampus, retrosplenial cortex, prefrontal cortex, and hippocampus." "An allocentric representation emerges from nonadditive computations shared across multiple interacting brain regions." In 2014, when Li, Karnath, and Rorden reviewed this complex topic, they also concluded that specific allocentric defects do exist in patients and that they often coexist with egocentric biases.[29]

Comment

What the normal brain's two spatial reference frames accomplish in milliseconds is still taking neuroscientists decades to sort out. Pathological lesions that also disconnect its networks have not made this sorting any easier. In another decade or so, the literature on this Self–other topic will become more coherent.

Meanwhile, our medial cortical networks seem to be asking and answering two Self/other questions: (1) *Who* am I (more anteriorly)? (2) *Where* have I been, and *when* (more posteriorly)? Many of these dialogues depend upon medial temporal lobe functions.

6

Early Distinctions between Self and Other, Focal and Global, Are Coded in the Medial Temporal Lobe

> Investigators are now attempting to understand the differences in functional properties among the hippocampal subfields and among the different structures that send afferents to the hippocampus.
>
> J. Knierim and colleagues[1]

Consider the single-cell amoeba. Before it ventures a pseudopod out into the external world, it has to consult some organic, self-discriminating information in its own outer membrane. If it couldn't sense where its own "inside" stops, it couldn't tell which food particle in the "outside" environment was safe to engulf and digest.

Evolution took off from there. Human brains now contain billions of interactive nerve cells linked together into big networks. Even so, this huge "Connectome" must still solve that same basic problem: how to preserve our own vulnerable Self's assets safely *inside*, while selecting those few morsels for focal interest that lie *outside* in that huge external volume of global space beyond the surface of our own skin.

In 2014, John O'Keefe, Edvard Moser, and May-Britt Moser shared the Nobel Prize in Physiology or Medicine. Four decades earlier, O'Keefe's laboratory had first reported the original research on hippocampal "place cells." These few place cells were only in a rat. Why were they so special? Because they provided any freely running rat with a "cognitive, or spatial map of its environment."

Pause and consider how crucial it was to have your very own "map" of 3-D space "online" ever since you got out of bed this morning! A recent review, coauthored by O'Keefe,[2] explains how different spatial nerve cells accomplish this remarkable feat of spatial cognition.

The world took more notice in 1991 when these place recognition cells were finally recognized in the primate hippocampus.[3] Again, these cells' firing responses were context-sensitive. They corresponded with the actual *place* the monkey occupied at that moment. Of these 21 neurons, 5 gave responses interpreted as egocentric; 6 gave responses interpreted as allocentric.

Head Direction Cells

Another category of neurons was also discovered that fired in response to spatial information. Researchers had linked these responses with neurons they labeled "head direction cells." [ZB: 487–492; ZBR: 101–108, 323–325][4] In the previous chapter, we found that seemingly sharp ego- and allo- concepts could get blurred when neurologists tried to apply them to gross lesions that damaged human brains. Could discrete recordings from single monkey neurons make these clinical issues easier to understand?

Which Direction? Whose Head Is Facing What? What Is a Compass "Heading?"

Words conceal ambiguities. It is easy to become confused. For example:

- The properties of "head direction" cells might represent a kind of *path integration*. Path integration refers to one way that the brain estimates its current *location and direction*, solely by integrating its own *internal* sources of data. Which internal data? Chiefly those subtle, private, sensorimotor messages it receives from its *own* proprioceptive and vestibular systems, and from the actual total motor output that also flows out to move its *own* muscles.

- On the other hand, "head direction" cells might derive their current position and orientation solely on the basis of different sources of *external* data (visual, auditory, tactile). These stimuli are extracted from the world *outside*. These local external cues could then be subtly processed by "*landmark* navigation."[5] Notice that these two approaches are not mutually exclusive. Each mechanism could, *and does*, influence a nerve cell's preferred firing direction.

Robertson et al. used video tracking to follow the movements of two monkeys under freely moving conditions.[6] They observed the firing patterns of five head direction cells. Four were in the presubiculum; one was in the parahippocampal gyrus. These cells conveyed much more information about head *direction* than they did about which external scene was then being viewed in the space outside, or about the actual location of the place the monkey was in, or about the actual position of that monkey's eyes. The authors provided an apt summary of their own impression: "One can think of the direction cells as responding like a compass attached to the top of the head which will signal head direction even when the compass is in different locations, including in a totally different, and even novel, spatial environment." These cells made intelligent decisions. They had gathered together sufficient information to "know," in a sense, in which *direction* of space "*their* behavioral North" might be located. Moreover, this particular knowing could then help the monkey *move* in a *goal-directed* manner toward a useful goal. [ZBR: 107–108]

Cells manifesting this sense of direction get off to a very early "head start" in life. Bjerknes et al. found head direction cells in infant rats as soon as 11 days after they were born.[7] Most electrodes were recording from cells in the subiculum (or in one instance, from the medial entorhinal cortex). These cells were already competent three to four days *before* the pups had even opened their eyelids. This kind of larger, hard-wired network is a handy, built-in system to have, especially when you're trying blindly, on frail legs, to scramble back to your nest in the dark. Head direction cells in adult rats also maintain their tuning for direction during complete darkness. However, their "memory" for this preferred tuning direction may fail if too much time elapses. Notice that subsequent visual input can help anchor the preferring firing direction of these cells to particular "landmark" items and events in the outside world.

Head direction cells are widely distributed. Substantial numbers exist not only in the hippocampus and hypothalamus but also in the anterior and lateral dorsal thalamic nucleus. [ZBR: 106–108, 172–174] Notably, they also occur up in the *anterior* part of the retrosplenial cortex, the region discussed in relation to landmarks and maps in the previous chapter. Therefore, head direction cells seem poised to co-sponsor vital lower-level and higher-level cognitive functions in subtle ways that can escape notice.

To the degree that these direction cells in the limbic system, thalamus, and retrosplenial cortex are tuned to respond to "spatial landmark cues *out there*," one could consider their broader functions as falling within the category of responses to "*otherness*." In contrast, to the degree that head direction cells represent proprioceptive, vestibular, or muscle movement mechanisms, their "*en*tuitive" sense of direction seems tuned to correlate with our *internal*, somatic, "*Selfness*" properties. Given such potential ambiguities, some researchers believe that we humans will continue to maintain our sovereign, biased gradient of greater covert Self-referent processing (i.e., ego >>> allo) *even* while we are attending to an item out there in so-called object-centered (allo) space. [ZBR: 532, note 5]

Commentary

Do our head direction cells represent only one simple underlying mechanism? Are they limited to only one minor role? Their properties, substantial numbers, and locations suggest otherwise. How could such distributed functions be viewed as playing a guiding role in behavior? By needling, nudging, and reshaping all those basic Self/other options that select, *subconsciously*, the most appropriate conditioned behaviors for us at pivotal points all along the Path we follow in daily life (chapter 12).

Grid Cells

Grid cells are different. They were first discovered in the medial entorhinal cortex and in the pre- and parasubiculum. [SI: 259–260] "Grid cells fire whenever the animal's position coincides with the vertices of a periodic triangular grid."[8] Although grid cells in a foraging rat initially fire in relation to very localized cues, with more prolonged experience they later fire in *global* coherent patterns that are consistent with large-scale global navigation.[9] Why can grid cells help in dead reckoning? Because they can fire independently of the context of landmarks.

Grid cells are intermingled with both head-direction cells and with cells intermediate between these two categories in the layers of this *medial* entorhinal cortex. Here, the grid cells keep "expressing variable degrees of a directional modulation." In contrast, the head direction cells keep "expressing variable degrees of grid structure." Moreover, the firing of all of these cell types is modulated by an animal's running speed (velocity). This serves as one more example of the reasons why researchers increasingly soften their distinctions by describing grid cells' functional properties as having been blended into "conjunctions."

Several caveats apply to these sentences, and to the magical term "conjunctions." They warn every reader that when adjacent modules interact, their functions resist being split into neat psychophysiological packages (appendix G). Words in English are, in the Zen phrase, "like a finger pointing at the moon." They can point toward the existence of neural codes, but they do not necessarily translate them accurately nor decode them. Thus forewarned, we begin the next section by pointing toward two parts near the front end of the long parahippocampal gyrus. Other attributes of this gyrus are further oversimplified in chapter 15.

The Medial and Lateral Entorhinal Cortex

The entorhinal region can be seen on the right side of figure 6.1. This figure places the entorhinal region in that short,

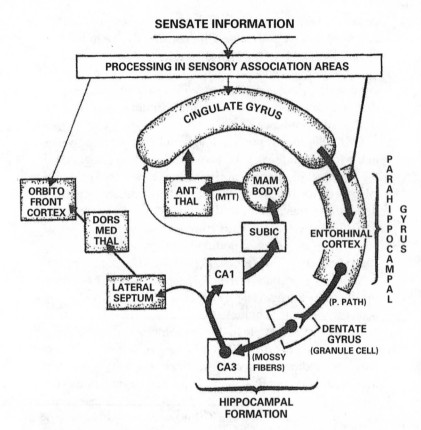

Figure 6.1
A hippocampal crossroad and the limbic circuitry. Impulses already undergo much processing on their way down to CA 3 cells (bottom square). From CA 3 cells, they are shunted on to CA 1 cells, next to the subiculum, and then relay to the mammillary body of the hypothalamus via the mammillothalmic tract (MTT). From the anterior thalamic nucleus impulses flow up to the cingulate gyrus and then back down to the entorhinal cortex of the parahippocampal gyrus. From there, they enter the dentate gyrus of the hippocampal formation via the perforant path (P. PATH) and relay on to the hippocampal CA 3 cells. This completes the Papez circuit, shown in thicker black arrows.

The CA 3 cells also send off a second branch. This path leads through the lateral septum on to the mediodorsal nucleus of the thalamus and most anterior portion of the parahippocampal gyrus as this gyrus nears the hippocampal formation.

The next figure (figure 6.2) outlines some major distinctions—and some interconnecting links—between the lateral and medial entorhinal systems as research suggests they exist chiefly in the medial temporal cortex of the rat. Like other oversimplified attempts to diagram the hippocampus and parahippocampus using arrows, this circuitry also represents a first approximation.[10]

As we have already emphasized, our brain has *two* different attention-processing pathways, dorsal and ventral (chapter 5). It would be reasonable to propose, therefore, that some links could exist between the *dorsal* processing pathway (*Where* is it in relation to *Me*?) and the *medial* entorhinal cortex. This proposition would then leave room for other links to occur between the *ventral* attention pathway (*What* is it?) and the *lateral* entorhinal cortex. Figure 6.2 illustrates the possibilities that such relationships exist between the MEC and the LEC. Unfortunately, whether these word-notions are first modeled on the findings in rats or monkeys, they don't yet encompass the whole story when we try to extend them toward the semantic complexities encoded in the human equation.

To illustrate, Knierim and colleagues conclude that the *lateral* entorhinal cortex (at left) is likely to process *local* cues—either "individual items or conjunctions of items."

thence up to the orbitofrontal cortex. Axons called mossy fibers relay impulses from granule cells to CA 3 cells. Many axons of the perforant path release glutamate, an excitatory amino acid (appendix B). The arrows shown on this and the next wiring diagrams are oversimplifications. Sometimes the situation is more reminiscent of a haystack. Countless pathways go in many—and frequently reciprocal—directions. The diagram is therefore highly selective and schematic. (Adapted and modified from multiple sources.)

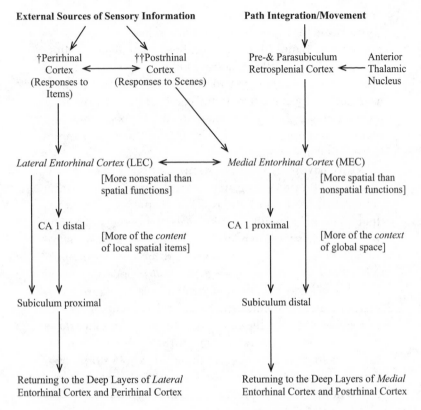

External Sources of Sensory Information

†Perirhinal Cortex (Responses to Items) ⟷ ††Postrhinal Cortex (Responses to Scenes)

Path Integration/Movement

Pre-& Parasubiculum Retrosplenial Cortex ⟵ Anterior Thalamic Nucleus

Lateral Entorhinal Cortex (LEC) ⟷ *Medial Entorhinal Cortex* (MEC)

[More nonspatial than spatial functions]

[More spatial than nonspatial functions]

CA 1 distal

[More of the *content* of local spatial items]

CA 1 proximal

[More of the *context* of global space]

Subiculum proximal

Subiculum distal

Returning to the Deep Layers of *Lateral* Entorhinal Cortex and Perirhinal Cortex

Returning to the Deep Layers of *Medial* Entorhinal Cortex and Postrhinal Cortex

Figure 6.2

Two pathways from the parahippocampus into the hippocampus via the perirhinal and entorhinal cortex: A schematic overview. Much of this figure is based on Knierim et al. (n.1) and on box 2 of Moser et al. (n.4). Many other pathways leading into and out of these regions cannot be included in so simple a diagram (n.10). Moreover, the lateral and the medial entorhinal systems differ in other ways. For example, the LEC innervates the *outer* third of the molecular layer of the dentate gyrus. The MEC innervates the *middle* third.

†In humans, the *perirhinal* cortex is heavily interconnected with the amygdala, the ventral temporopolar cortex, and the lateral orbitofrontal cortex (see n.12). Its activations correlate both with our sense of familiarity and with our learning of meaningful associations about objects, including their affective and motivational significance. ††In rats, the *postrhinal* cortex responds to scenes in a manner similar to the ways our parahippocampal gyrus responds to scenes.

This would enable the LEC "to provide the content of an experience, including the spatial information related to these local objects." Despite some ambiguities, these words sound almost as though the most *lateral* cortex (the LEC) had the greater capacity to process local external "landmarks" (appendix G).

In contrast, the *medial* entorhinal cortex "is best thought of as part of a global, holistic spatial map." Yet this global map "is generated primarily through path integration mechanisms." These integrations "provide information about where the organism is in its environment, where it is going, and how to get there." The word integration sounds as though the most *medial* cortex (the MEC) could be more capable of monitoring "inside" data, as discussed earlier in this chapter.

Semantic issues arise. How *do* you and I code for all of *our* own space *inside* and for all of that *other* world of space *outside* and then blend the two? We keep returning to these topics of Nobel Prize caliber—ego/allo, focal/global, nonspatial/spatial—in parts III and IV. Don't be surprised to find that you'll be invited to take new perspectives on old figure/ground relationships.

Some Potential Implications of the Medial and Lateral Pathways in Figure 6.2

The right upper side of figure 6.2 shows that some input enters from the anterior nucleus of the thalamus. These early messages can be transmitted not only down into the subiculum but also upward into the region of the retrosplenial cortex and posterior cingulate. So, two questions can arise about the phrase "spatial *context*" as its usage relates to one's personal Self along this MEC pathway. For example: "*Where* is it (this goal) in relation to *Me*? *Where* am *I* going?"

Parenthetically, in *kensho-satori*, no such Self-referential questions of context remain in the foreground or background of consciousness. All sense of one's agency in "*doing*" drops out. No sense remains of any forward movement toward any goal. [ZB: 538, 543] (Speculation: might this strengthen the possibility that the reticular nucleus had blocked some of this earlier input from entering into the MEC, say, by inhibiting it at the level of the dorsal tier of the thalamus?).

In contrast, the normal lateral pathway (LEC) has seemed more likely to be in touch with the word, *content*, as it applies to local items within a scene. Perhaps it could also highlight certain discrete ongoing external events. The hybrid interconnectivities of the entorhinal cortex in figure 6.2 illustrate why this figure could leave unstated any *simple* attempt to identify the LEC exclusively with "*What* is out there?" (a question of local *content*) or could pause before identifying the MEC exclusively with "*Where* is it? This question relates global *context* back to the axis of Self (appendix G).

However, one plausible interpretation occurs with further regard to *kensho-satori*. It is that many of our normal processing functions elsewhere are also organized and interconnected in a similar manner. This could also enable them to briefly manifest that same striking shift of properties—identified here symbolically as allo >>>> ego—in ways that express the basic phenomenology of such states. During these major extraordinary alternate states of consciousness, the GABA "cap" of the reticular nucleus of the thalamus could inhibit those multiple anterior and dorsal early contributions that lead toward such medial pathways. This kind of targeting of the GABA inhibitory process could still spare those other more ventral contributions that lead into the allo- functions of the lateral pathways. [ZBH: 24, 118–119]

Another caveat applies to our estimates of the sense of "time" in laboratory animals. When a trained rat runs on a treadmill,[11] the distance (and the elapsed time) that register in its grid cells are not easily translateable into an explanation for the profound vacancy of "time" that can enter the experience of a human subject during *kensho*. [ZBR: 372–383; SI: 193–196; ZBH: 59–61, 190, 192]

High Resolution fMRI Studies of the Entorhinal Cortex in Humans

Two separate groups, in Germany and the Netherlands, have used high-resolution fMRI (at 7 tesla) to clarify the wiring diagrams of the LEC and the MEC of normal human subjects. In general, their results confirm the outline in figure 6.2. For example, the perirhinal cortex preferentially connects with the anterior LEC. The parahippocampal cortex is preferentially connected with the posterior MEC.[12]

In humans the entorhinal cortex is small, measuring only 25–30 mm.[13] However, the posterior MEC is most sensitive to scenes (as is our parahippocampus). It is also the most connected to such posterior regions as the occipital and posterior parietal cortex. In contrast, the anterior LEC is most sensitive to items-objects, and it is most connected to such anterior regions as the medial prefrontal and orbitofrontal cortex.[14] Furthermore, the posterior entorhinal cortex is more responsive to spatial scenes than to nonspatial objects. The anterior entorhinal cortex is more responsive to nonspatial objects-items than to spatial scenes. Notably, our *intra*-entorhinal connections were also found to cross over the boundary between the LEC and MEC. This confirms the potential for cross-talk to blend our various physiological functions.

Hippocampal Outflow Systems

Aggleton's review[15] summarizes these outflow pathways that contribute to different aspects of our memory. The first three are these:

- The "reciprocal hippocampal-parahippocampal system." This is involved in our sensory processing and integration, as just discussed. Its connections with the entorhinal cortex flow in both anterior-posterior and medial-lateral directions (figure 6.2).

- The "extended hippocampal system." This services our *episodic* memory. It proceeds via the mammillary body to the anterior thalamic nucleus, cingulate gyrus, and retrosplenial cortex.

- The "rostral hippocampal system." This pathway contributes to our affective and social learning skills. It proceeds via the prefrontal cortex, amygdala, and nucleus accumbens.

Comment: A Grounding in the Actual Daily Life Events of Human Subjects

How do I remember in which room I *mis*placed my glasses? There's still more to learn about how the human brain lays down and retrieves all of its codes for location, direction, navigation and procedural memories. [SI: 72–76, 259–260] We'd also like to know why the LEC is most vulnerable to the inroads of Alzheimer disease. Functional MRI is hobbled by a seconds-long lagtime. This pace is too slow to clarify with millisecond precision exactly how the synaptic sequences in my entorhinal—hippocampal cortex interact with those in my precuneus and retrosplenial cortex in ways that direct me to "head" in some direction (yet with no clear mental image in mind) where I will actually *find* those glasses that I'm searching for. [SI: 72–74]

Recent research by Chadwick and colleagues at University College, London, concluded that "Both direction and location codes are present in the human retrosplenial cortex, and they are anchored to the local environment through salient cues."[16] Moreover, they then monitored 16 subjects with fMRI while they were making directional judgments about a particular goal. These judgments required the subjects to remember previous information that they had laid down earlier. At that time, they had been moving around freely in a simple, ground-level virtual environment.[17]

Signals in the left entorhinal cortex and subicular region were equally active whether the subjects were simply facing in a given *environmental* direction and/or whether they were representing the direction of an intended goal. The evidence suggests a way to resolve some potential semantic ambiguities between those two words, context and content: "Head-direction populations are intially involved in representing current facing direction." They "then switch to simulation during navigational planning." This same entorhinal—subicular region gave no indication that it represented egocentric directions. In contrast, the subjects' *left precuneus* was the most active region during navigational coding for this kind of *egocentric* information. On the other hand, it showed *no* evidence that it was representing environmental (geocentric) information.

Whether we are navigating life's actual twists and turns to the bathroom in the pitch dark or trying to follow straight arrows that hint at crossing pathways through the human brain, it's good to pause and be reminded: small local activations in diverse parts of the brain can join together instantly to become large, unified "conjunctions." You'll discover this practical principle illustrated on the green side of each dollar bill. Find the bald eagle that holds 13 arrows together in its left talons. Just above it is the Latin phrase, *E pluribus unum*.

With this reminder of the precarious unity of a fledgling nation, we turn next in part III to other contributions made by our living memory systems. They help us generate more than a long, coherent personal history. They accomplish more than automatic short-term working memory skills. Some even hint that we posess archaic instinctual skills. Neuroscience still gropes to define these, let alone measure them.

Part III
Aspects of Memory

Great is the force of memory, excessively great, O my God. Whoever sounded the bottom of this large and boundless chamber?

Saint Augustine (354–430)

Remindfulness

There is a guidance for each of us, and by lowly listening we shall hear the right word.

R. W. Emerson (1803–1882)[1]

Gratitude is not only the greatest of virtues, but the parent of all others.

Marcus Cicero (106 BCE–43 BCE)

Mindfulness has gone "viral" in recent decades and has expanded into a movement.

J. Wilks and colleagues[2]

Being Mindful of Sati

The word *mindfulness* has taken on many different shades of meaning. At the core of most current practices is the cultivation of an attitude that attends to each present moment, *non-*judgmentally, at some level of conscious awareness.

This interpretation alone, useful as it is, leaves out an important ingredient. This second attribute was inherent in the ancient meanings of *sati* in the original *Sati*patthana Sutra. Indeed, for the Dalai Lama today, "The most important meaning of mindfulness is recollection."[3] So, perhaps it's time to pause and remind ourselves about the links to memory in this old Pali word. Originally, *sati* had key links with the brain's *memory retrieval* functions. [MS: 92–106] These memory *retrieval* functions are crucial. They enable us to access a given event that had occurred in some [prior] present moment and then recall it "back to mind at a *later* time."[4] Notice what this "later" recollection implies. It means that *automatic* memory systems—with no conscious effort—retrieve such a moment from our *long*-term memory stores many *decades* after it had first been stored there. [MS: 156–163] Moreover, these recollections can become the foundation for that vital, everyday practice of gratitude. Cicero considered gratitude the parent of all virtues. [MS: 140–145]

Viewed in its broader context, *sati* includes our capacities to *access subconscious, intuitive memory functions*. Precisely how can long-term meditative training contribute to these vital, *spontaneous, supraordinate* processes of retrieval? Over the decades, does something intrinsic to authentic meditation help cultivate remindful links among our covert associations? Can this "something" help ripen a critical subliminal mass of useful *insights*? Can these insights help us finally change our overconditioned, maladaptive behavior? [ZBH: 75–78]

Starting Small: Just Following the Breath Mindfully

What about counting the breaths in and out? You might first think that this is a task too simple and too boring to help anyone on a so-called "spiritual path." It's not. [ZBR: 58–61; SI: 225–227] Levinson and colleagues at the University of Wisconsin recently confirmed the fact that following the breath has highly practical aspects.[5] Indeed, their detailed studies establish that our competence in counting the breath is a useful behavioral index of our capacities for mindfulness.

First they needed to select one operational definition for mindfulness. It was the conventional definition. Almost everyone could agree that it is *present moment awareness*. Then, in four classical experiments (on over 400 subjects), they arrived at these conclusions:

- Accurate breath counting correlated with greater degrees of meta-awareness, with the trait estimates of mindfulness, and with fewer mind-wandering episodes. It did not happen to correlate with one of the ways they estimated their subjects' capacities for working memory.

- Accurate breath counting correlated with *lesser* degrees of "wanting." [ZBR: 264–265] For example, competent subjects were less "captured" by the same colors that they had used in previously rewarded tasks.

- In meditators, accurate breath counting correlated with how long they had been practicing and with their lesser tendencies to be engaged in mind-wandering.

- When subjects were trained for only 4 weeks in breath-counting tasks, they showed fewer mind-wandering episodes. The subjects involved in control tasks did not.

Standard Tests of the Working Memory Network

Normal adults use a common core network for recollection when they solve three standard memory tasks.[6] fMRI monitoring shows that these memory regions include the left angular gyrus, the medial prefrontal cortex, the hippocampus/parahippocampus, the posterior cingulate cortex, as well as the left middle temporal gyrus. Notice that better performances during these recollection tasks were associated with increased connectivities between the two frontoparietal and cingulo-opercular *attention* network systems. [ZBH: 79–80] The authors appreciated that such increases in connectivity (which might seem to link covarying attention modules high up in the cortex) did not exclude this possibility: both sets of signals might originate in a central "pacemaker" down in the thalamus. [ZB: 404]

What Is a Potential Role for the Hippocampus in Normal People Who are More "Mindful" and in Long-Term Meditators?

Start with certain normal, *non*meditating young adults who *are* more mindful of the present moment. Do their hippocampal gray matter volumes differ from those other normal controls who are *not* more mindful? The 247 subjects (averaging 21.7 years in age) were studied at Beijing Normal University.[7] The trait of mindfulness was based on *self*-reports. This trait was estimated in each group using the Mindful Attention Awareness Scale (MAAS). Modest correlations emerged. Students who *self*-reported that they tended to be more mindful of the present moment had larger gray matter volumes in their right hippocampus/amygdala and in their anterior cingulate cortex on both sides. On the other hand, their gray matter volumes were smaller in the posterior cingulate cortex bilaterally and in the left orbital frontal cortex. No gender differences or correlations with general intelligence were evident.

Did the brains of *long-term* meditation practitioners differ from those of age-matched controls?[8] Both groups contained 30 subjects who averaged 47 years in age. These meditators had a very long average duration of practice: 20 years. Structural MRI studies showed that the meditators' hippocampal dimensions were increased over those of their controls. Both male and female meditators had larger left and right hippocampal volumes. However, in males, the *left* hippocampus had the larger radial distance. In females, the *right* hippocampus showed the larger radial distance.

A separate report compared the amount of atrophic change in fifty long-term meditators and fifty matched controls. In the meditators the left hippocampus/amygdala and left posterior cingulate gyrus were included in the cluster of regions that showed the *least* decline in volume between the ages of 27 and 77.[9]

Chapters 1 and 2 indicate that a major long-range emphasis in Zen is to become *more* insightful.[10] So, let's be clear: insights of different kinds, sizes, and practicality arrive at different times on the Path, and they strike in different settings. [SI: 125–129] Off toward one end of this spectrum are our ordinary superficial hunches. Off at the other end are the *extraordinary*, existential insights. The outlier is the exemplary state that had awakened the Buddha under the Bodhi tree. However, such advanced degrees of *selfless* enlightenment strike suddenly. During the states of *kensho-satori*, their comprehensions open instantly into the *direct* experience of *insight-wisdom* (chapter 2). *Prajna* is the Sanskrit term for this lightning flash of deep realization. [ZB: 545–549 (J. *Hannya*)]

Many lesser insights can ripen during the course of a *balanced* long-term practice, one that blends both concentrative and receptive forms of meditation. Some of these remindful, lesser "openings" help solve our simpler, everyday kinds of *working* memory tasks. They also help resolve a wide variety of life's ordinary difficult issues, the kinds that require longer incubation times to be reflected upon and reappraised.

We use our ongoing subconscious surveillance mechanisms to monitor these ordinary events. Operating in a Self-correcting "intelligent" manner, these ordinary *involuntary* capacities tend to keep our best short-term and long-term interests in mind. As meditators gradually hone all their attentive and awareness skills, they can begin to benefit increasingly from developing a clear, objective, subtle, *overview* awareness.

Let's now return to examine how things happen during Zen meditation (*zazen*) as our focused attention keeps gently monitoring each breath, in and out. Of course, attention soon lapses, and our mind starts wandering. Wandering is its natural discursive tendency. In retrospect, three hidden mechanisms of our basic, subconscious *remindful* capacity might then seem reflexly to leap into operation.

First, without *our* trying, one mechanism detects this lapse into mind wandering. A second mechanism now disengages our thought stream from each discursive distraction. A third mechanism gently *re*-collects attention, and instantly *re*directs it back to its original intended focus. [ZBH: 78–81] A short version of these three working memory sequences would be:

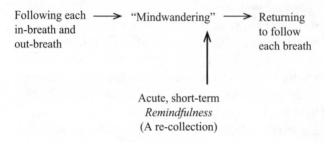

Following each ⟶ "Mindwandering" ⟶ Returning
in-breath and   to follow
out-breath   each breath

Acute, short-term
Remindfulness
(A re-collection)

Long-Term Remindfulness: A Momentous Historical Example

One spring, when young Prince Siddhartha was still a child, he happened to meditate under a rose-apple tree. This incident became his first discovery that *meditation, under a tree, could bring him into a blissful state of consciousness.* [ZBH: 33–40] Almost three decades later, we find him exhausted from his rigorous six-year spiritual quest. He feels he can't go on. Now this pivotal event from far back in his childhood springs to mind. It *retrieves itself* from his long-term subconscious memory stores. Where will this crucial recollection lead him? Fortunately, once again, it leads him to seek out a tree under which he can meditate. This tree under which he finally became fully awakened will be known thereafter as the Bodhi tree (*Ficus religiosa*).

Remindfulness

No dictionary contains the word *remindfulness*. It can be simply defined as *the quality of recollecting the most appropriate thing at the right time.* The remarkable adaptive powers of such normal, involuntary affirmative memory skills have long been recognized. Emerson viewed some of these subconscious, Self-correcting skills as our natural, intuitive source of "guidance." As each small insight arrives, with *no* word-thoughts attached, we often find it striking with utmost clarity and authenticity. Various reminders have the power to guide us in practical ways that are more effective than ruminations or pangs of conscience. Some can lead us to make much better tactical and strategic decisions than we would ordinarily. Still bigger decisions can lead toward career-changing life choices that express our most deeply principled, affirmative aspirations (chapter 25).

Remindfulness begins with the opening of a *subconscious* form of receptivity. When cast involuntarily, such a wide net can reach far and wide into memory storage sites, enclosing whatever it contacts. [ZBH: 148–149] By way of an example, let's return to expand on that earlier short list where only three working memory sequences might have seemed to operate during ordinary Zen meditation:

- In quiet meditation, *one* part of our working memory stands watch. High in the crow's nest it remains poised, on silent standby alert. Its job is to *monitor* and signal relevant internal and external events. (Table 1.1 suggests that *both* major styles of meditation involve monitoring.)

- This barest awareness detects—sooner or later—that we're drifting away, in toward the shoals or out into the open seascape of mind-wandering. That gap between our earlier good intention and this reality has now grown too wide.

- The instinctual guidance aspects of remindfulness decide to correct this mismatch.

- Signals alert the helmsman, calling for a swift skillful recovery back to our intended course.

- Steered back by these next automatic shifts of the rudder, we rediscover ourselves again following each breath, in and out.

We've now identified *five*, not three, silent subconscious sequences. These five stand ready to service our needs for Self-correction. Separate, millisecond, *monitoring* responses are easy to overlook and hard to measure. What's the simplest way to think about these reflexive shifts? *We can think of them as long-term training and retraining exercises.* [SI: 22–29] For every meditator on a cushion, each such trial-and-error step can serve as a kind of *mini-insight*, returning the brain toward its original intention. Notice that innumerable course corrections, retrainings year after year, could gradually establish, and then help reinforce, a deep, habitual *attitude* of ongoing *re*mindfulness.

While meditators are off their cushions, they confront the vicissitudes of everyday life with wide open eyes and ears. Again, similar silent recollections keep recurring, especially during pauses. Suppose a new problem arises. To solve it means sifting through decades of long-term memories, accessing widespread association networks, forging new links (chapter 1). [ZBH: 177–181] Many siftings and shiftings are likely to engage the habitual operations of those deep, silent networks that link the striatum and thalamus with the cerebral cortex. [MS: 133–139]

Out on the job, complex ethical decisions must be made. Sophisticated social assessments are required in order to adjust one's interpersonal behavior (chapter 15). On other occasions, perhaps while driving a car at highway speed, one needs to access different long-term *procedural* memory skills. Fortunately, this range of practical, preconscious, mini-insightful solutions leaps silently from the parallel processing networks distributed throughout our brains (chapter 19). [MS: 148–149, 171]

The paragraphs above sample only a few of our dynamic, reflexive capacities for surveillance, detection, and reaction. These monitoring procedures shift "intelligently," detaching instantly from the present moment and shifting into memory retrieval. Do all these dynamic sequences unfold too fast to be measured by fMRI? During meditation, fMRI reveals that the phase of mismatch (which signals the mind-wandering gap) can be correlated with the activations of certain cortical regions in both hemispheres. These are the anterior/middle insula and the dorsal part of the anterior cingulate cortex.[11] [ZBH: 78–80] We await the kind of millisecond monitoring with MEG that can optimally resolve such short intervals (appendix C).

Meanwhile, does solid experimental evidence indicate that certain forms of creative problem solving proceed subconsciously?

Silent, Spontaneous Modes of Implicit Processing

Some refer to this implicit processing as "unconscious thought."[12] On these pages, the word, *subconscious* seems preferable. Moreover, why does *processing* serve our need rather than "thought?" It is because, during this processing, no "thought" per se is being heard internally. [MS: 145–155]

Recent articles aired some controversies in this field.[13] On balance, the existing evidence suggests that silent modes of implicit preconscious processing do exist. However, the fact that they arise spontaneously makes demonstrating them a delicate matter. Investigators often introduce crucially different conditions into their experimental designs. The resulting literature is so uneven that a well-intentioned "meta-analysis" is prone to be led astray despite its statistical assets. Can we really expect subconscious processes to work better when burdened by *major* distractions? Can the accuracy of all decisions be inferred, often in student populations, by tests that use different automobiles as consumer products? The applicability of such memory-invoked value judgments to authentic Buddhist meditation techniques that emphasize a *silent* pause for reflection is open to question.[14]

Recent Studies of Implicit Processing

Abadie et al.[15] began by drawing a distinction between our two forms of memory encoding: (1) *verbatim* representations: these are exact, detailed concrete representations of the basic facts; (2) *gist* representations: these represent the general meaning of information and our more abstract contextual associations to it. The researchers then conducted two behavioral experiments on 97 university students. The tasks did focus on a down-to-earth issue: which was their best choice among four different apartments?

The results indicated that (1) evidence supporting the so-called "unconscious-thought effect" (UTE) did correlate with an enhanced memory for the *gist* of those attributes that were relevant to the decision-making process; (2) the subjects' enhanced memory for the *gist* of decision-relevant attributes did coincide with their actual improvements in decision making. However, the improvements occurred *only* if the distraction task used was relatively low in its demand characteristics. This study met the criteria for common sense: a test situation that would *not* impose so big a multitask during a delicate mode of covert mental processing that it would create a major distraction.

Vlassova and colleagues[16] found in a behavioral study that subconscious information accumulated slowly. Thereafter it could be used *only if* it had access to certain kinds of information that were relevant to the decision-making process. Again, although the accuracy of their subconscious processing was enhanced, these subjects were *not aware* of the fact. This *lack* of confidence in having made the correct decision is evidence in favor of implicit processing. [SI: 153–158]

Li et al.[17] tested whether subconscious processing made it easier to detect certain rules underlying the way two types of letter strings had been arranged into patterns. Subjects who processed the strings subconsciously were more accurate in arriving at decisions based on grammatical rules.

Zhou et al.[18] found that inherent differences in cognitive styles affected their subjects' decisions about four different types of phones. Subjects who were prone to use holistic cognitive styles were equally good at separating the good from the bad phones whether the task involved conscious or subconscious processing. In contrast, the subjects who were prone to use analytical cognitive styles performed most accurately only when they went on to use subconscious forms of processing.

Vallee-Tourangeau et al.[19] focused on the use of playing cards. They studied a large number of subjects of diverse ages during five different experiments. The subjects' success in using statistical modes of reasoning emerged more or less spontaneously. (This kind of reasoning required no added instruction.) However, their greatest successes in statistical reasoning occurred when they could actually *handle* the cards. Now they could manipulate the layout of the cards "in the real world" rather than just "in their heads."

This last study illustrates the merits of a "hands on," actual *living* approach to problem solving as opposed to just an intellectual approach. Thinking *about* Zen is never intended to replace *the actual direct experience of Living Zen in daily life.*

Can Problem Solving Involve Parts of Our Brain That Are Relatively Far "Out of the Loops" of Language?

We've seen that people learn how to extract regular statistical probabilities from sequences of events without being aware that they have done so. This subconscious feat is called implicit probabilistic sequence learning (IPSL).[20] Performance scores on such ISPL tasks are predictable from one's resting-state fMRI connectivity map. Higher ISPL performance scores (without awareness) correlate with *greater* connectivity between two regions: the dorsal caudate nucleus (R, L) and the *right* medial temporal lobe. The higher ISPL scores correlate with *less* evidence of connectivity linking this dorsal caudate with those premotor cortical regions that are involved in motor planning. This result fits in nicely with the previous evidence that shows that *destructuring* one's prior planning sets is the crucial prelude to breakthroughs into novel insights. [SI: 169–173] A recent report indicates that variations in the activity of the left caudate nucleus tend to be predictive of positive thoughts about one's own personal past.[21] Chapter 1 describes other recent research that implicates the caudate nucleus in intuitive processing.

Chai et al.[22] monitored normal children, adolescents, and young adults with fMRI while they were studying various scenes. In all age groups the hippocampus was more activated by those original scenes that proved later to have been remembered most successfully. On the other hand, it was those subjects in the *adult* group who showed the most *deac*tivation of their so-called default network by the original scenes that were later remembered. In contrast, children showed the least reliable deactivation of this same network by scenes that they later remembered. The data can be interpreted to suggest that humans learn to deploy greater attentiveness skills during their long processes of development.

Indeed, it will be our trainable, inconspicuous attentiveness and awareness skills (plus subtle mechanisms still to be defined) that also enter into the task-induced *de*activation of the default mode network.[23] [SI: 75–76, 266; MS: 189]

Why do we originally attend to certain items and events, and more deeply encode them? Is it because they seem to convey a special sense of meaning in general or appear particularly relevant to us as individuals? [SI: 130–146] These deeply encoded salient episodes are often better remembered and tend to have a greater motivational impact.[24] Yet it is still difficult to untangle all of the mechanisms responsible for each such difference between deep and shallower levels of attentive processing. Some of this difficulty could begin with the fact that researchers are not yet measuring all of the many linkages, direct and indirect, that bind the processing of attention and awareness to the many mechanisms involved in memory. [MS: 154–155]

The next chapter continues to explore the diverse brain pathways that help us remember.

8

A Remindful Route through the Nucleus Reuniens

The nucleus reuniens distributes massively to the hippocampus and in a highly organized manner.

R. Vertes and colleagues[1]

Unconscious encoding in the human depends on the hippocampal-anterior thalamic axis and its connections to the neocortex.

S. Duss and colleagues[2]

Decades of research have informed us where multiple facets of memory become organized—chiefly down in our medial temporal lobe, especially in the hippocampus and the para-hippocampus that funnels into it (see figures 6.1 and 6.2). [ZB: 180–189, figure 6; ZBR: 99–108; SI: 80, 96–97, 107, 259; MS: 119; ZBH: 93, 116]

Projections from the subiculum are the major output pathway for impulses issuing from the hippocampal CA 1 cells.[3] We've seen that these hippocampal messages first descend to the hypothalamus. They then speed up from its mammillary body to the anterior thalamus (using a special white matter tract). Then, from this anterior thalamus, they next relay up to the anterior cingulate gyrus. From here, some can reach as far back as the deep medial posterior parietal region that includes the posterior cingulate gyrus and the retrosplenial cortex. As impulses feed forward and feed back throughout these memory circuits, their reverberations generate multiple associations in our other brain regions that extend beyond these initial examples.

For instance, consider the huge amounts of personal history you've amassed and how much you can recollect in reasonably intimate detail. How could any three-pound lump of tissue keep encoding such an ongoing Self-referential journal, let alone maintain this narrative and retrieve information from it? [SI: 70–71] Recent research focuses on several connections that link the thalamus with the cortex.[4] We might specify these bidirectional networks (at least by labeling them with words) as the thalamo ↔ prefrontal, thalamo ↔ posterior cingulate, and thalamo ↔ retrosplenial networks. [SI: 88, plate 4]

In this quest to trace our autobiographical roots, let us now start with the associations of our highly evolved *pre-frontal* cortex. Are its medial networks "smart" enough to be in selective two-way communications with the hippocampal memory functions down in our medial temporal lobe? Can they participate in deciding that *this* immediate event is *really* worth remembering, that *this* past event *really* must be recalled? And, how could we ever know (at some conscious level or subconsciously) that our MPFC had actually been sending and receiving such signals?[5]

A singular fact: this vital signal from the prefrontal cortex has no direct, *non*stop route down into our hippocampus. It must pause to make *one* stop. Only after this relay can these messages reach into the hippocampus. Pause *where*? In the *nucleus reuniens* of the thalamus. [MS: 109, 217]

The Reuniens

The nucleus reuniens is the only *midline* nucleus among the several nuclei that make up the "limbic thalamus." All three other nuclei are higher up in the *dorsal* thalamus. [ZBR: 171–174; SI: 90–92] The simplified diagram on the next page helps appreciate why this midline nucleus, so close to the *medial* temporal lobe, becomes a contributing member of the limbic thalamus.

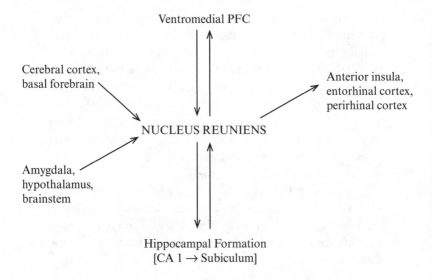

Normally, the reuniens appears to exert a direct, potent excitatory influence both on CA 1 and the entorhinal cortex. In addition to this, the excitatory drive from the reuniens might also exert a dual (+,–) effect on hippocampal CA 1 cells that can modulate their output through the subiculum.[6]

Recent localized lesions of the reuniens in rats suggest that it could normally contribute to functions loosely describable as a kind of "prefrontal executive inhibition."[7] Lesioned rats were more efficient than sham controls in certain prefrontal tasks. They showed enhanced focused visual attention and visual associative learning, together with reduced premature impulsive responses. Other experiments in intact rats suggest that the reuniens could normally enter more into aspects of memory consolidation and into remote memory retrieval than into the acquisition of new memories.[8,9]

Who Needs a Midline Thalamic Nucleus? For What?

Taking the millions-of-years evolutionary perspective, nuclei occupying the midline of the brain tend to have been cast in simpler, but essential, physiological roles. Although the human reuniens measures only 14.6 mm in length, it is poised at a crucial junction. What might a nucleus contribute that links our medial prefrontal cortex with our hippocampal formation? Could it be some fast, automatic, "intelligent" decisions that helped ancestral species survive with the aid of short-term, long-term, even instinct-based memories?

Recent fMRI studies of subconscious relational memory inferences are especially relevant. Zust et al. monitored 33 normal male subjects.[10] First, the researchers presented them with subliminal (masked) stimuli. These visual stimuli combined unfamiliar faces with written occupations. Notice that these stimuli were much too brief for their contents to fully enter consciousness. Subsequently, the subjects saw the same faces, but this time these faces were presented *supra*liminally. The question now was: Could the researchers measure the degree to which those *previously* masked faces had served as covert cues? Did those earlier subliminal traces of faces still help reactivate the subjects' *subconsciously* linked written occupations? The fMRI data suggested that the hippocampus did have a role in the subjects' *initial subconscious* encoding and retrieval functions, not just in the way they might *consciously* encode and then later retrieve these face/occupation associations. The results favored the interpretation that some overlapping of networks *within* the hippocampus itself could enable a previously activated but faint memory trace to later trigger the reactivation of the subsequent memory trace.

Networks high[11] and low[12] contribute to our normal subconscious memory configurations, not just those in both hippocampi. Notably, the various modules that were coactivated in this human study of subliminal faces also included the *ventromedial thalamus,* not just those bilateral contributions expected from the amygdala, superior temporal sulcus, superior and middle temporal gyrus, and the temporopolar cortex. How do all these regions normally interact? What happens when some drop out? These are among the topics explored in the next chapters.

9

A Disorder Called Transient Global Amnesia

> The contribution of CA 1 neurons in the human hippocampus to the retrieval of episodic autobiographical memories remains elusive.
>
> T. Bartsch and colleagues[1]

On the Varieties of Normal Memory

It helps to review and condense what happens normally in the vast scope of our memory before we turn to what causes this memory to drop out. Put simply, we first *encode* an episode into memory, then *consolidate,* and finally, *recall* it. [ZB: 259–262] Other words refer to this sequential process as the "laying down" of memories, their "storage," and their subsequent "retrieval" or recollection.

Memories are of several different kinds. A different network system in our brain sponsors each kind. *Working* memory systems help us to monitor and solve a new task immediately, in this present moment or in the very short-term. *Long-term* memories are different. They're already of two different kinds before we even start to consolidate them into memory "storage sites."

> 1. *Explicit Memories.* These draw on facts that we can *declare* verbally or in writing. We're more or less consciously aware of having these facts. An example: white feathers cover a mature bald eagle's head. (Could the word-label for this bird "ring a memory bell" that was first "struck" back in chapter 6?)

Explicit memories help us clearly recall single facts, episodes, or larger events. We've seen that they are referable initially to processes chiefly in the medial temporal lobe—to the hippocampus/parahippocampus and their many connections.

We register explicit memories in the usual, ongoing *anterograde* manner. This means that they start with events in *this* present moment and then continue to be laid down during each pending future moment. *Retrograde* memories, once registered, enter a time-based gradient of storage. They keep on getting older. We might later retrieve them from a time far back in the remote past of our childhood. Autobiographical memories are those that document our perceptions, feelings, and impressions of events in our own personal past. They reinforce our sense of a historical Self that is ongoing.

> 2. *Implicit Memories.* These involve automated behaviors that express our habitual *procedural* memory skills. They reflect the ways that our behavior has been conditioned subconsciously in the course of long experience. Your handwritten signature is one example.

Some memories can register *sub*consciously, subliminally, without our being aware of them. Memories can be forgotten or repressed, temporarily or permanently. Memories can be accurate (true) or inaccurate (false). They are also prone to be modified in the process of being retrieved. [ZBH: 99–123] With this preamble, we come to a distinctive, relatively benign neurological disorder of memory.

The Disorder Called Transient Global Amnesia

Certain middle-aged and elderly patients suddenly experience this acute, severe memory loss.[2] This brief disorder of ongoing, anterograde memory wipes out their immediate memories as fast as each new event presents itself. Thus, in the emergency room, the patients repeatedly ask: "Where *am* I? … What day is it? … What year is it?" Even when you supply the correct answers repeatedly, they cannot retain this information in working memory for more than a few seconds. Most patients remain alert, talkative, and capable of high-level language functions. Notice the uneven memory loss: they are usually aware of *who* they are but not *where* they are. Unfortunately, not all cases follow this classical clinical picture. Fortunately, they usually regain their memory after several hours. [ZB: 188]

The disorder is slightly more common in women. Notably, it often occurs in relation to stress, excitement, straining on a closed glottis, or other dynamic physical activity, including intercourse. Transient global amnesia (TGA) can be accompanied by a lesser degree of retrograde memory loss. This means that some events are deleted that happened just before the anterograde amnesia started.

In the past, when typical, mild cases were accurately diagnosed in ordinary clinical practice, they tended not to be overly investigated aggressively.[3] However, during the last five decades, the dramatic nature of this severe memory loss, the arrival of the latest neuroimaging techniques, the occasional presence of venous asymmetries,[4,5,6] and the fact that other lesions can produce a similar syndrome,[7,8] have all stimulated research efforts to understand why this disorder occurs. Why do these new neuroimaging findings in TGA take on added significance? Because their sensitivities have a bearing on our potential to study at least the "afterglow" of other acute states of consciousness. These states include *kensho-satori*.[9]

Recent Neuroimaging Research

Among these newer magnetic resonance imaging (MRI) techniques, several diffusion-weighted imaging (DWI) methods have proven informative (appendix D). Note that the recommended approach is to use a slice thickness of only 2–3 mm, *and* to perform the second DWI study *two or three days after the acute onset of TGA symptoms*.[10] In the classical benign form of transient global amnesia, these *delayed* images can often identify small, focal, dot-like hyperintense lesions in the hippocampus, *even after the patients appear to have regained their previous lost memory*.

For example, Bartsch et al. studied one large group of 16 TGA patients with diffusion-weighted imaging at the University of Kiel, Germany.[11] Their acute transient attacks of amnesia had lasted for an average of eight hours. Every patient in this study had small, focal, dot-like *lateral* lesions on the outer left or right sides of their hippocampus.[12] Figure 3 in this report illustrates, in three dimensions, that all lesions were laterally placed. Figure 4 indicates that one particular portion of CA 1 was affected: the region called Sommer Sector. [ZB: 182–184]

A recent study from Israel by Peer and colleagues followed 12 patients during the acute phase of their transient global amnesia.[13] None had focal neurological signs, epilepsy, or a recent head trauma. These researchers used both fMRI during the "resting state" and voxel-based morphometry (VBM) to detect the resulting structural changes. TGA had recurred in 2 of the 12 patients.

Notice how early all the patients in this group had been scanned by diffusion-weighted imaging. They were all scanned *less than* 14 hours after the clinical onset of their amnesia. Only one patient, who happened to be scanned only five hours after the onset, showed a hippocampal lesion when diffusion-weighted imaging was performed *this soon*. These observations emphasize how important it is to *perform a second DWI two to three days after the onset*, allowing time for these delayed DWI changes to evolve.

Fortunately, five patients consented to have a second, *resting* state fMRI scan. This was performed two to nine *months after* their acute episode. Therefore, this report did include some assessment of longitudinal changes in brain connectivity. The hippocampus was one seed region serving to represent the connectivity within the modules of the "episodic memory network." The left angular gyrus was chosen as another seed region representing the "language network." The seed region for the "motor network" was in the precentral gyrus. The amygdala, medial prefrontal cortex, and the cingulate cortex (anterior, posterior) were regions considered to represent the "stress-related network."

Using these fMRI seeds, the results of the changes in connectivity were intriguing:

- Functional connectivity was more disturbed in *both* hemispheres during the first hyperacute phase of transient global amnesia than it was when the post-acute fMRI scan was performed several months later.

- This bi-hemispheric disturbance in functional connectivity was widespread. It was specific not only for the hippocampus and the "episodic memory network." It also included the other networks cited above.

- Connectivity was reduced not only between the hippocampal and these other network clusters. It was also reduced *among* those connections of the episodic memory network that linked each of *its* parts with the others.

- Scans in the post-recovery period did not otherwise differ significantly from the scans of healthy control subjects.

Comment

These DWI and fMRI studies suggest that both the early-phase *and* the later-phase *longitudinal* investigations are essential when investigating TGA and relevant acute states of consciousness. Please notice: TGA is *not* presented as an exact model of *kensho-satori*. Many phenomena in these two states are diametrically different. Their neuroimaging topographies will surely differ at key points. Rather, TGA illustrates some proactive principles essential to test a set of plausible working hypotheses: (1) Organize a research team months in *advance* of an event, (2) Design a protocol that includes both appropriately targeted diffusion-weighted and fMRI, psychophysiological and biochemical measurements, and first- and second-person reports, (3) Continue to perform the relevant DWI and fMRI studies, searching for three kinds of changes: Those that take time to evolve during the first few days; those that subsequently disappear; and those that could persist. Each type serves to bring the requisite *longitudinal* dimension that helps to clarify which mechanisms are involved in producing the state.

Words Don't Accurately Describe Coded Dynamic Functions

Encoding, consolidation, recall—each of these standard terms used to introduce chapter 6 is a glib condensation. Bartsch and Arzy[14] review the current literature's facile tendencies to use such short-hand word labels in trying to tag separate parts of the normal hippocampus with different normal cognitive functions, for example, (1) The dentate gyrus—with "pattern separation"; (2) the CA 3 network—"with pattern completion"; (3) Both dentate gyrus and CA 3—"with memory encoding" and "*early* retrieval"; (4) The CA 1 network—with the huge burden of "*late* retrieval, consolidation, recognition, autobiographical memory retrieval, novelty, mismatch detection, and map-like representations of an environment."

Bartsch et al. recently reviewed the various cellular pathogenic mechanisms that can selectively damage CA 1 neurons *and* can then cause them to die within the next 48 to 72 hours.[15] Notably, *this* particular group of TGA patients was now considered to have "no clouding of consciousness or loss of personal identity." They knew *who* they were. Thirty percent of these 53 TGA patients gave a history of migraine. [ZBR: 306–312]

More remains to be said in appendix D about the potential sensitivity of DWI techniques to study the mechanisms of *kensho-satori*. More is said about word problems in appendix G. More still needs to be added about memory and the medial temporal lobe. These additions can be found both in the notes to this chapter (notes 1, 2, 3, 12, 13, and 15), in the remaining chapters of part III, and in chapter 17.

10

Remindful Zen: An Auditory "Altar Ego"?

Memory is the treasury and guardian of all things.
Marcus Cicero (106–43 BCE)

Our brain stores a lifetime of memories. A rare few slip back into our awareness at odd times, taking the form of hallucinations. Hallucinations are peculiar, unguarded sensory perceptions. They do seem real, but no appropriate external stimulus explains them.

For the past 12 years, every two months or so, I've experienced a certain kind of brief auditory hallucination. This benign form recurs only in the early morning hours. I hear it only when I'm in light sleep, in bed, usually between 5:30 and 6:15 a.m.—never when I'm wide awake. Therefore, this curious sensory perception rides on the rising tide of my morning arousal into wakefulness. By convention, the term *hypnopompic* is used to describe these hallucinations that surf on the crest of one's *ascent* from sleep toward waking consciousness. [ZB: 340, 381–386] (Hypnagogic hallucinations occur while one is *descending* into sleep.)

To Freud, our slips of speech were glimpses into the operations of our unconscious. We call these misarticulations "slips of the tongue." They seem to issue more from the active, motor side of our psyche. In contrast, benign auditory hallucinations are clues that issue more directly from the *sensory* aspects of our psyche. These perceptual slips also allow us to glimpse subconscious levels of our sensory processing. The things that we can hear at such levels slip into awareness only if our association networks have free access to a particular kind of "lowly listening."

Don't confuse the minor, benign events recounted in this chapter with the kinds of major hallucinations that have a more serious import. [ZB: 30–34, 374] I present them here as clinical observations, together with these kinds of qualified reassurance: (1) hallucinations can be expected to happen on the meditative Path; (2) they are phenomena consistent with "mind-manifesting" quickenings that tend to arise during various physiological transitions [ZB: 371–465];[1, 2] (3) they are not harbingers of disease and disaster to be avoided at all costs; (4) still, they can become temporary diversions from the essential Buddhist themes of insight-wisdom and compassion.

Meanwhile, why describe the three most recent phenomena? Because, as a matter of scientific interest, they illustrate the refinements of a nuanced remindful function that might serve a useful purpose. These incidents linked the emergence of a novel, practical, auditory repertoire with a definite increase in the frequency of my Zen meditation practices—indoors *and* outdoors. Clearly, repeated periods of mediation have acted as a catalyst for brief, remindful quickenings. [MS: 156–163]

Single, Soft, Bell Rings

It began with a single *soft* note: one ring of the familiar doorbell. On two occasions, convinced, I got up and went to the door. No one was there. Strange. Later it was usually one ring of the telephone. Again, only a single note.

No bell ring occurred in the context of any dream. Instead, these single rings interrupted my light sleep near the time when I would be about to awaken anyway. Occasionally, the sound was a single, short "cling" of the electric alarm clock. So then I wondered: Could some kind of electrical short-circuit in the house set off these bells? Later, the sound became the single note of the *hand*-wound alarm clock. No wiring in the house could account for *this* clock's ring. Could such recurrent sounds arise spontaneously in my brain?

The neurologist finally accepted the obvious facts. Yes, these single note rings were all recurrent auditory hallucinations. When any bell ring was lateralized, it arose in the space vaguely off to my left. I never found my head turning reflexively to orient toward the apparent source of any bell (chapter 12).

The first benign auditory hallucination occurred during that period of turmoil associated with the final two years of my wife's Alzheimer dementia. However, the later hallucinations were clearly related to different factors. For example, most brief rings had more immediate preludes. They occurred much more frequently when (1) I had been meditating more frequently, and for longer periods, (2) My everyday wakeful level of awareness was higher and clearer, and (3) I was also in good physical shape after having exercised the day before doing vigorous outdoor work, hiking, or playing tennis. During the next 12 years, as most of these same mild episodes kept recurring at their original frequency, I accepted them as just another personal idiosyncrasy on the Path of Zen. [ZBH: 124–143]

Something else was gathering momentum. In recent years, if the rings occurred somewhat earlier, say around 4:30 a.m., I found that this was when my bladder was reminding me that it was time to be attended to. Ringings of this kind introduced a new possibility: some of these stereotypical single bell rings were timed in a way that might also serve a subtle utilitarian function as a urological "wake-up" call.

Then came the three most recent reminders. Each recollection illustrated the emergence of a novel auditory repertoire. Nothing like these recollections had ever surfaced before. [ZB: 395–397]

First Episode: A New Kind of Bell with Two Novel Reminders

It is May 8, 2015. Early on this Friday morning, I'm still in a phase of light sleep. Abruptly, I hear a single soft resonant note. This gently struck bell sounds vaguely off to my left.

Had I heard *this* dulcet bell before? Yes, countless times during the past nine years. This was the regular meditation bell at *Hokoku-an* zendo. Meditating there, earlier this same week, I'd heard this real bell rung three times — not only on this past Monday, and again this Tuesday, but also yesterday. Starting at what time? At 5:30 *p.m.* Each cluster of these real bell ringings resounded in the late *afternoon*. In fact, three evenly spaced soft strikings of this same bell were the customary signal that would have opened each of those three extra, late afternoon, half-hour sittings in the zendo.

Now, back home in bed in the dark, I'm slightly aroused by this bell sound, just enough to recognize it as a new, minor variation on those other old, familiar bell hallucinations.* But then a quick glimpse at the clock reveals that it's still only 6:00 a.m. *Much* too early to get up. Plenty of time to slip back to sleep, where I had just left off. So I drop back into sleep once again …

Maybe 20 minutes later, I'm still sleeping. Now I hear two short, single-note *whistles*, off to the left. The first note is higher than the next. Two seconds later, this pair of whistled notes is followed by a short "*Psst!*" Neither auditory event is recognizable as a sound that *I* had produced in the past or in the present. As I arose from bed and started walking, these three new events were easily recognized, and their sequences were readily interpretable:

- The first two whistled notes sufficed to serve as an abbreviated alert. Why were these first two notes special? They were the start of the Austin family's seven-note signal call. I have whistled, and heard whistled, this whole string of seven home-grown signal notes ever since I was a small child. These two notes were recalled from retrograde memories, *way way* back.

- This "*Psst!*" was different. It was a sophisticated nudge. The intent to reinforce was subtle, yet obvious. The purpose of this

---

* Over the decades, I've heard countless bells open and close each meditation period. Yet, *never* had I hallucinated any *mediation* bell, anywhere, before or since.

novel sound was clearly to remind the laggard sleeper: that first bell was your *initial* wakeup call. It had been ignored. Now, it's *really* time you got out of bed!

• A single bell ring is a relatively simple phenomenon, a brief sound retrieved passively from memory storage. Whistles enter from a different operational level. Humans form whistles; bells can't. A whistler needs to purse lips. Thereafter, even to create an audible "*Psst!*" requires a very complicated kind of human investment in intention, in pucker and articulation simply to enable this sound to be heard and recognized. This silent sound was no simple, ordinary, auditory hallucination!

Commentary

Never before, during the late afternoon of any week, had I supplemented my regular, daily routine (30 minutes of before-breakfast meditation) with *three additional* half-hours of before-supper meditation in the zendo. So, the first extra days of this week were exceptional: already a new total of five hours of meditation on the cushion. The extra hour and a half of meditation appeared to have helped sponsor the emergence of these three novel auditory events: meditation bell, whistled signal notes, "*Psst!*"

Never before had I heard any two-note abbreviation of our family whistle signal. Never before had a warning "*Psst!*" also arrived to jog this laggard sleeper into finally rising from his bed. Not in *my* usual vocabulary was such a "*Psst!*" used to attract someone else's attention. On the other hand, suppose I *had* wanted to quietly alert a family member. Then these first two notes and this "*Psst*" would have been the simplest, quickest, best way I could have ever chosen to do so.

We are now searching for words to describe the presence of an unusual kind and degree of human behavior. This behavior, expressed in a benign hallucination, is manifesting itself creatively at some rarefied level of spontaneous cognition. It had just inserted itself into a sleeper's "ordinary" 12-year, bell-ring repertoire. Had an almost "friendly agency" deftly condensed a seven-note family signal into only two notes? If so, it was some agency with a sense of tact, one that could issue a well-meaning, urgent reminder to this sleeper to finally get up from bed without further delay.

This whole episode provided a glimpse into a hidden autonomous capacity for guidance. It had the ability to monitor behavior in a time-sensitive manner. It could discern the gap between intention and reality, issue a signal, and act responsibly *in anticipation of* a sleeping person's best interests.

Second Episode: "By the Light" ... A Syncopated Refrain

It is three weeks later, May 29, another Friday morning. Again, I'm sleeping about 5:45 a.m. All those routine, daily prebreakfast meditations had taken place as usual on the four earlier mornings of this week. Yesterday (Thursday), I had occasionally hummed and whistled the usual opening note—in a *slow* tempo—to the melody of the old song, "By the Light of the Silvery Moon."[3] They went like this:

THURSDAY

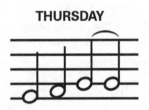

Now, sleeping, I hear three *rising* notes, whistled off to the left. These whistles seem on the same pitch—*E, F, G*—as those three notes for "By the light" that start the song I'd whistled *yesterday*. But this morning's version sounds different. Its first two whistled notes are in a *fast* tempo, lively, *syncopated*. These two new notes have bounce! This tune had been corrected overnight, during my sleep, without my conscious participation. These dotted quarter and eighth notes no longer dragged!

FRIDAY A.M.

Commentary

You might also have noticed something else: *No extra* sitting meditation had taken place this week from Monday through Thursday. So, could anything else have served as the prelude to the almost "executive-level" capacity of this early Friday morning's musical hallucination? Yes. That day before was highly exceptional. In retrospect, it was unlike any other day I'd spent during the past 10 years.[4]

That Thursday, yesterday, in midafternoon, I had been the sole beneficiary of a silent, *solitary*, inspiring, hour-long Nature walk. [MS: 59–60; ZBH: 33–48] The trails on this 22-acre Audubon sanctuary led through a mature forest of magnificent old, tall trees and lush, fresh green foliage. This woodland sanctuary was a dedicated space for outdoor walking meditation and for re-creation. Yesterday afternoon, even the birds were observing silence. Scientific studies document the fact that such natural sylvan settings benefit both brain and body (appendix A). [ZBH: 186–189]

The ancient Taoists and Ch'an masters understood that Nature exerted a major influence on persons who followed a spiritual Path. Master Yongming Yanshou (904–975) phrased it as follows:

> Mountains and rivers, earth and grasses, trees and forests— these are always emanating their subtle and precious sound. Day and night they demonstrate the ultimate truth to each of us. It's right in front of our face. Everyone has this inconceivable capacity for great liberation.[5]

Third Episode: Two Flute Notes after a Previous Indoor and Outdoor Prelude

It is early Sunday morning, November 15, 2015. I'm asleep about 5 a.m. I hear two short, soft, airy flute notes. Both notes are on the same medium pitch. They are referable (first time ever) to the space off to my *right*. They are followed by a soft, dissonant noise that lasts only a second.

Why these distinctive *flute* notes? Only after several hours would I be reminded of their origin. Three weeks ago, I had heard the haunting timbre of music played on a Japanese bamboo flute (*shakuhachi*). This solo performance lasted for only five minutes or so. It was an impromptu, incidental part of the entertainment program at a Science and Non-Duality Conference.

Once again, I had spent three extra half-hours on Monday, Tuesday, and Thursday meditating in the zendo between 5:30 and 6:00 p.m. (in addition to my regular morning zazen). Moreover, late that Saturday afternoon (the 14th) I had taken a rare 45-minute Nature walk down another wooded trail. Tall trees lined each side, but by then most of their leaves had already turned color and fallen.

Background Information: Normally, What Kinds of Internal Sounds Do People Usually Hear?

We hear a lot of our own inner spoken thoughts. Some of this *inner speech* is *deliberate*. How does this Self-cuing function solve simpler problems? By helping us to retrieve useful associations and to explore better options for representing a pending goal.[6]

These more *voluntary* forms of verbal thought activate many brain regions. We use them routinely when we perform other concrete, goal-directed acts of *attentive* processing. These almost intentional thoughts are in a different category from most of our more *involuntary* mind-wandering and unuttered wordy thoughts. We hear the latter, spontaneous discursions more typically when we are in the so-called "resting states" during times of external silence. Their Self-referent content tends to be more abstract in its associations. During fMRI monitoring, involuntary thoughts correlate with activations in those standard regions of the default mode network that can also include the lateral temporal cortex, hippocampus, and parahippocampus.

Perrone-Bertolotti et al.[7] recently reviewed the last half-century of (often controversial) research into this *inner speech* that is sometimes called "the little voice inside our head." They estimate that this pervasive internal language occupies at least one quarter of our conscious waking life. In many people, inner speech buzzes like a bee hive. For centuries, Zen has offered an antidote. This approach to meditation means an explicit emphasis on *no word-thoughts*. This will gradually allow intrusive emotional reverberations to drop off. [ZB: 293; SI: 150–152; MS: 145–150; ZBH: 155–156]

Auditory hallucinations of speech are a special kind of word thoughts. [ZB: 395–397] The lowest prevalence rate for auditory hallucinations in the general population was a mere 0.7 percent. It was obtained when researchers asked this specific question: "Did you at any time hear voices during the past year saying quite a few words or sentences when there was no one around that might account for it?"[8] (We'll discover later that other estimates can be as much as 100 *times* higher.)

The fMRI connectivity data obtained during *auditory verbal hallucinations* were recently reviewed.[9] The results were mixed and often contradictory. Some evidence suggested that connectivity is disorganized in the left temporal lobe in general and in the left superior temporal gyrus in particular. These findings could reflect (1) an imbalance in temporal lobe glutamate and GABA mechanisms or (2) a reduction in the inhibitory tone from the frontal lobe.[10] Other contributory factors are worth further investigation.

For example, normal subjects can be entrained to react to a rhythmic sound stimulus. This phenomenon is referred to as the auditory steady-state response.[11] One current theory is that this 40 cps auditory steady-state gamma response serves as a useful index of one kind of excitatory transmission. Enhanced responses, "driven" by excitatory amino acids, could underlie the heightened temporal lobe perceptual sensitivities that develop in meditators. This theory could be tested during rigorous meditative retreats.

Affective resonances from the limbic and paralimbic systems generate the emotions that modulate what we normally hear. These varying emotions infuse notions of pitch, intensity, tempo and voice quality. When normal subjects hear such cues, fMRI signals in their auditory association cortex and amygdala appear to be relayed into the inferior frontal gyrus (IFG), where they could undergo further evaluation.[12]

Researchers can use transcranial magnetic stimulation (TMS) to transiently disrupt the normal right parietal cortex.[13] [ZBH: 115–116, 238 n. 15] The results confirm that the right hemisphere is the more specialized for directing our immediate attention to higher-order auditory events that exist out in left hemi-space and *global* space. [SI: 31]

Lowly Listening

When Emerson introduced the phrase, "lowly listening," he was referring to a subtle form of wise guidance (chapter 7). [MS: 145–147] "Lowly" does not mean that only the lower levels of our brain would be contributing every necessary ingredient. In fact, networks high and low feed back and forth normally in the process of integrating the requisite information (chapter 18). This means that top-down supervisory functions continually monitor and adjust our sensitivities to fresh perceptions. When doing so, expectant networks first scan some likely options, and then recall—*selectively*—only those relevant experiences that had been coded minutes to years before. When a clinician discloses a first-person report about the kinds of unusual auditory hallucinations described above, the scientific purpose is straightforward: contemporary research can benefit by reminders that we still have much to learn about how (seemingly unusual) brain mechanisms (currently obscure) can accomplish such extraordinary integrations and recollections of memory.

Lowly listenings seem to attend to both our internal *and* external global landscapes. Their subconscious, supervisory, scanning activities have tacit permission to access skillful means (Skt: *upaya*). Having thus detected a mismatch of events, they seem ready to sponsor an executive role that issues off-stage cues, *sotto voce*. These remindful receptive functions resemble those of some invisible guardian—reminder—coach. It's easy to overlook such insightful communiqués. In my case, an aged brain has issued only brief, prosaic messages. The sensory data condensed in these heralds of morning arousal had been laid down hours, weeks, decades earlier. No high-level, career-changing decisions have been suggested (as would happen during the auditory events that the wild parrot protector describes in chapter 25). Still, notice how such small, ordinary reminders might serve a practical memory role in everyday life.

Disclaimer

Nothing esoteric, mystical, or supernatural is implied in this account of benign auditory hallucinations. They are recounted solely to illustrate the dynamic range of subconscious neural resources that a reasonably normal long-term senior meditator's brain might tap into. When? During discrete surges of early morning arousal activity. [ZB: 457–460] These are hints of a forthcoming reveille. They reflect this person's individual cyclic, endogenous brain-body rhythms, not his immediate, overt reactions to an actual external sensory stimulus.

The evidence points to two prior conditions that favor their arrival: (1) more frequent meditation on the cushion indoors and (2) leisurely walks outdoors among tall trees. [ZBH: 186–189] The *delayed* hallucinations that emerge only in the early morning—long hours after these earlier conditions had reinforced each other—might seem to be "little openings" or "little quickenings." [ZB: 371–465]

Provisional Nomenclature, Scientific and Vernacular

Which words best describe these subjectively *real*, brief, elusive, supervisory/advisory functions? Let's begin with their singular *early morning* onset. Hypnopompic is a technical term for events that strike on the threshold of a person's arousal toward wakefulness. Decades before, this witness had experienced the *visual* hallucination of a maple leaf. Retrieved from the past, it heralded an internal absorption (chapter 16). [ZB: 376, 390, 470–471] Now, the events were all *auditory* hallucinations, also lateralizing mostly to the left side of space. Once again, these correlated with the increased frequency of meditation on the cushion supplemented by meditation practices in the outdoors.

Benign? After 12 years, almost certainly.[14] Frequency? A recent World Health Organization report estimated that the lifetime prevalence of auditory hallucinations was only 2.5 percent.[15] In other recent surveys of the population at large, the frequency of auditory verbal hallucinations reached 7.3 percent.[16] Among healthy college students, as many as 71 percent would report auditory verbal hallucinations while they were awake, and 39 percent reported hearing their own thoughts spoken aloud.[17] When nurses and nursing students were surveyed, as many as 84 percent(!) described hearing voices.

Lewis-Hanna et al.[18] identified 12 healthy subjects who had auditory hallucinations either when they were on the verge of waking up (hypnopompic) or falling asleep (hypnagogic). They then gave these normal subjects the task of *deliberately* directing their *auditory* attention toward a number. During this auditory focusing task, these subjects activated their anterior cingulate gyrus more so than did their controls. [SI: 20, 23, 172] This observation could fit in with the suggestions that the two subjects to be described in chapters 11 and 25 were under some degrees of increased attentiveness and stress.

So, *benign auditory hypnopompic hallucinations* can serve as a provisional scientific term to identify a particular kind of little quickening that merits further investigation.

Yet what about the basic character of this present author's three recent auditory hallucinations? Obviously, they're *not* the voices referred to either in the reports just above or in chapter 25. Most of the recent ones are simple reminders—*sounds* retrieved from the past: a stored treasury of bell sounds, condensed musical notes.[19] But then came the syncopated whistles and the furtive *"Psst!"* These new developments suggest some kind of guardian agency that seems poised to do more than supervise. It can *act*, with discretion, on one's behalf. Its emergent repertoire hints at functions that provide the unwitting recipient with access both to a kind of "private eye" *and* "private ear." Moreover, as dawn approaches, these nuanced functions have the capacity to *act* tactfully. This is more than might be expected from the simplest kind of "inner voice" or "inner sound."

Recent recollections of this kind can be viewed in a *positive* light. Too often lately have we given our memory a bad rap. We've been thinking of memory in terms of negative ruminations, psychic trauma, PTSD, painful flashbacks, and guilt-ridden associations. It's time to emphasize those positive qualities of *remindfulness* that can help redress this imbalance.

Could some short, vernacular wording suggest such an *affirmative* "presence"? A subtle presence perhaps slightly more accessible in the case of a senior meditator who was already on a decades-long "spiritual" Path? Earlier centuries had floated notions of an invisible, friendly "genie," one that could exercise mysterious beneficial powers. Today, perhaps for people in a lighter, playful mode, "altar ego" might seem to be one possible candidate. Altar could convey the notion that whatever "presence" of mind might exist, it could function in ways that were more substantial than ethereal. Indeed, if any such underlying elevated structure were to be raised on an imaginary transept, it would seem to represent a sturdier, more pragmatic phenomenon than an exalted concept lying forever beyond reach.

The play on the words, alter ego, serves as a reminder: these guidance functions resemble those that might be served by a kind of trusted second Self. Ego keeps the phrase grounded. It specifies that such an altar ego is a resource of neural origin. Subtle mechanisms are waiting for researchers to discover in a recipient's *brain*,[20] rather than in some metaphysical belief system of extra neural origin.

11

Following an Auditory Stimulus, Then "Seeing the Light"

We all know what light is, but it is not easy to *tell* what it is.
Samuel Johnson (1709–1784)

Visual hallucinations arising in the context of meditation practice may serve as indicators that homeostatic plasticity has been activated and that the brain may be more malleable to learn and change.
J. Lindahl and colleagues[1]

E.G. is a 52-year-old man.[2] When he was 28, while meditating, he had a dramatic experience of seeing light. However, the unusual *auditory* trigger is the point of interest in this account. It was only after he first paid rapt attention to this unexpected auditory stimulus, as it *moved overhead*, that he experienced the light and opened up into a brief, alternate state of consciousness.

His routine practice at that time was to meditate for 20 to 25 minutes each evening in a darkened room at home. Under *closed* eyelids, he would gaze upward at an angle usually around 30 degrees above the horizontal. On occasion, he might feel "drawn" to gaze up farther in this vertical gaze position.

While meditating one evening, he suddenly heard the sound of a jet airplane high in the sky. "The sound came from behind me, back on my right side. It then moved overhead, and disappeared off to the left side in front." This was an unusual event (his home was not under the flight path used by any airport). He followed this sound as it moved overhead, from back to front, for 30 seconds or so until it subsided. This jet sound prompted an unusual, urgent, conscious desire to *be up there*. "I heard no thoughts, but was conscious of wanting to be up where the plane was." He does not remember whether or not this particular desire translated into an actual gazing up higher than was his usual custom. It was clear, however, that his attention was being drawn to (and focused on) this moving sound and that he had developed some emotional investment in it. Old personal boundaries and distances were being effaced.

In this darkened room, and with his eyelids still closed, "a crack then gradually opened up in the upper right portion of my vision. It became a brilliant white light with a striking blue tint." This light enlarged to involve both upper visual quadrants, primarily in the region between 10 o'clock and 2 o'clock. The light always remained *above* his visual horizon. He bathed in the awe of this bright light for perhaps the next two to three minutes. Any sense of the passage of time, however, was not obvious.

Before the light arrived, he had been experiencing his usual ordinary ongoing sense of Self. However, during the light, this ordinary sense was no longer present. As the light waned, there occurred a gradual recovery of his usual sense of Self. Afterward, he recalled saying to his wife, who had been sleeping, "I have seen the light." He described the content of the experience in his journal. He believes that his interpretations included the expression: "There is no question, God *is*" … There is "some force greater than me" [sic].

While he found this experience of light to be greatly reas-suring, he remained puzzled by the deep question, "What do I *do* with this experience?" He gradually discovered that his previous job in advertising was no longer fulfilling. Per-haps in this respect, the whole episode of experiencing "the light," plus the questions it raised, could have initiated a gradual process of transformation. This process seemed to have opened up a very different, *socially oriented* career path: one of serving other people, caring especially for the disad-vantaged who most needed help. Within a year and a half he had chosen his present occupation—social service work.

Commentary

Meditators experience a wide variety of "light" phenomena. [ZB: 373–386] In the Zen tradition these are regarded as *ma-kyo* and are not encouraged. For example, when I mentioned to Nanrei Kobori-Roshi that I had seen a leaf, he admon-ished me, saying, "When you concentrate too hard, you may see things." [ZB: 472] On the other hand, converging reports suggest that "lights may, at least in some cases, be signs that the practitioner has obtained a certain degree of concentration."[3]

Whereas Zen traditions downplay hallucinations of light or sound as mere side effects of meditation, these lesser "quickenings" can also serve as useful probes. They help to identify the mechanisms by which meditative practices can produce their mind-manifesting effects. [ZB: 377–465] For example, the Surangama Sutra directed many Chinese and Japanese Buddhists' attention to the fact that sounds are the most powerful of all triggering stimuli. [ZBR: 39][4] What is so special about sounds? Is it only because they register up in the auditory cortex? Is it because this cortical region is so close to the "circuit-breaker functions" of the temporo-parieto-occipital junction (TPJ) and to the temporal lobe's remarkable capacities for interpreting meanings? [SI: 130–141; ZBH: 55–56]

Neural Networks for the Attentive Processing of Sounds

As a generalization, the *auditory* pathways in our cortex also follow the two what/where attention processing streams. The more *ventral*/anterior auditory pathway supports our *identification* of objects (such as jet airplanes) and speech.[5] The more *dorsal*/posterior auditory stream supports our Self-referential processing of both *where* things are heard in space (*location*) *and* how they are heard moving in space. Subject E.G.'s narrative indicates that this unexpected sound, recognized as a jet overhead, was activating *both* his "what" and "where" pathways. The auditory zone of the claustrum, with its extensive connections, could contribute in a general way to the salience of this new stimulus.[6] The prompt and sustained presence of the illumination in his *upper* visual fields clearly points to a substantial auditory triggering effect, one that spread to further activate his lower occipital → temporal lobe attentive processing pathways. [ZBH: 100–101]

By paying close auditory attention, normal subjects can detect when researchers make only slight changes in the tones of two simple notes. MEG monitoring indicates that their *right* hemisphere detects these tweaks.[7] They begin in the superior temporal region (at 100 ms), peak next in the superior temporal sulcus (at 130 ms), then in the posterior ventral frontal region (at 230 ms), and finally in the anterior temporal cortex (at 370 ms). Notably, after this *right* posterior superior temporal gyrus is resected, patients no longer can track *moving* auditory stimuli.[8]

Recent Research on Auditory Stimuli at Different Distances

X. Zhang et al.[9] monitored the fMRI responses of eight adults to nonmoving 500-millisecond sounds that originated in different parts of "near" 3-D acoustic space. Again we find that the *right* superior temporal gyrus was consistently activated in each of the following spatial localization tasks:

- Making distinctions between sounds in the left and right auditory fields: these activations involved the right midsuperior temporal gyrus and the left posterior superior temporal gyrus.

- Making distinctions between sounds in the superior and inferior auditory fields: these activations involved the left posterior middle temporal gyrus and the right superior temporal gyrus.

- Making distinctions between sounds in the anterior and posterior auditory fields: these activations involved the left midportion of the middle temporal gyrus and the right midportion of the superior temporal gyrus.

However, please notice what happens during many actual experiences in ordinary daily life. Our brains' responses are usually exposed to several different moving, auditory, visual, and *combinations* of audiovisual stimuli. Moreover, as chapter 13 indicates, these stimuli often enter and depart at distances that are hundreds or even thousands of *feet away from us*. For example, visual attention tracks the back of a train and keeps *seeing* it moving away; auditory attention *hears* this same train clatter gradually subsiding hundreds of feet away. In subject E.G., auditory attention alone kept tracking the sound of the moving jet airplane as this plane flew thousands of feet overhead. These examples illustrate that substantial distances are actually involved in *real* "far space." This distant outdoor domain is not being tested for in current laboratory experiments of what the reports are calling "extrapersonal space" (someplace beyond reach) or "far space."[10] [ZBR: 323–325]

In the report by Yue et al.[11] single or bimodal auditory and visual stimuli were presented at either 50 cm (1.6 feet; "near") or 250 cm (8.2 feet; "far"). In this instance, the evidence suggested that the visual stimulus effect predominated over the auditory effect, especially when tested in this "far" space.

Van der Stoep et al.[12] distinguished between single auditory, single visual, and combined *audiovisual* stimuli in space at two different distances. This "near" space was tested at 80 cm (2.6 feet). "Far" space was (still close) at 208 cm (6.8 feet). The experimental design was comprehensive. The authors controlled for how the size of the retinal image decreases, and how the auditory intensity also decreases when the stimuli originate at greater distances from the observer. Their conclusion: *greater* degrees of *combined* audiovisual *integration* did occur, even adjusting for these two conditions of smaller retinal images and less intense auditory stimulation. However, this mutually reinforcing effect created by the two stimuli came into play *only* when these auditory and visual stimuli were presented at the greater ("far") distance from the observer.

So it seems that innate, reflexive mechanisms could combine to enhance our attentive perception when factors of greater distance enter into the processing equation. Some of these early orienting mechanisms could arise in the leveraged amplification properties of the polymodal nerve cells found down in the colliculi of our midbrain (chapters 14 and 22). [ZB: 241–242]

"Hearing with One's Whole Being"

Ch'an master Yun-men Wen-yen (864–949) commented on 48 classical koans in the koan collection entitled *The Gateless Gate*.[13] In case 16 he referred to an intimate moment of major unification. Such a state of awakening can leave the impression of hearing with one's *whole being*, not just with the conventional faculty of hearing per se, nor with simple levels of absorption. Languages use different words when trying to describe the several phenomena during such extraordinary moments. English ventures the phrase "Just This," or "suchness." Chinese uses *Chi-mo* or *Shi-mo*. Japanese uses *In-mo*.[14] [ZB: 549–553] The words refer to the (word*less*) inexpressible, essential nonduality of Ultimate Reality. [SI: 11–13; ZBH: 8–9, 194–195]

From a neural perspective, the suchness of *kensho-satori* is Just This realization: the entire environment of forms, and the total emptiness of all prior overconditioned constructs of Self, are manifest simultaneously in One eternally perfect, immanent, Universal Reality.

The Color of the Light in Subject E.G.

The prominent blue hue of E.G.'s light is a point of interest. Human visual cortex is especially responsive to external blue colors. [ZBH: 139, 234 n. 24] "Sky gazing" is a form of meditative practice in Tibetan Buddhism. [MS: 89–91]

External and Internal Triggering Stimuli

Most triggering stimuli in Zen lore seem to have been prompted by external sources of stimuli. However, the movements of one's own body can also contribute to the critical mass of connectivities that culminate in *kensho-satori*. The next chapter discusses the subtleties of turning.

Turning

> The practice of Zen consists of a collection of liberative techniques that rest on a profound analysis of human perception and conditioning.
>
> J. Cleary and T. Cleary[1]

Zen practices point toward liberation. [ZB: 637–645] This means freedom *from* the old pejorative Self and freedom *for* its adaptive, affirmative, intuitive counterparts. [SI: 197–206] Some relatively simple retraining techniques tap into neural resources in ways that can be profoundly liberating. For example, suppose you turn suddenly, then stop after 90° or so. Now a whole new visual field greets you. In this act of turning, you also stimulate your semicircular canals. The canals relay some impulses instantly up to your insula. The posterior insula is an often overlooked part of your "vestibular cortex." [ZBR: 95–99; SI: 253–256]

Turning in Zen

In earlier centuries, "turning" had direct implications in Zen training. [ZBR: 39–40, 108–109] Baizang Huaihai was a Ch'an master who lived from 720 to 814. Like others in the vigorous Huang Bo-Linji line of masters that would follow him, Baizang engaged in abrupt, surprising behaviors.

Sometimes these happened soon after he had entered the lecture hall, and even before he gave his dharma talk.[2] On such an occasion, he might suddenly leap up and wield his staff as though to drive his monks from the hall. Then, just as they were running out of the hall, he would call out to them. When they turned around, he would demand, "What is it?" At other times, he would go on to deliver his lecture on the dharma. Then, while his monks were dispersing, he would call out to them. Again, as they were *turning* around, he would demand, "What is it?"[3]

Starting in the Tang Dynasty (618–907), the direct question *What*? became another teaching tool in the Zen repertoire. As an external stimulus, this word was intended not only to capture a monk's attention. It was designed to do so at a *pivotal* moment, *literally*. Later, in the Rinzai Zen school of eighteenth-century Japan, the question *What*? was also out at the leading edge of Master Hakuin's memorable koan: "What is the sound of one hand?" The same word was later revived in the context of twentieth-century neuroscience. Ungerleider and Mishkin introduced *What*? as the question that the *ventral* attention-processing system was expected to answer as impulses passed through the lower pathways in our temporal lobe.[4]

Standing Up

Different parts of our vestibular systems are stimulated whenever we move up or down in a vertical dimension. This movement subjects us to the effects of gravity that arise from the mass of the underlying Earth. I learned about some practical aspects of this phenomenon from my teacher Myokyoni, who had trained in the Kyoto school of Rinzai Zen.

During a Zen retreat in 1982, she invited us to observe what happens in our inner experience as soon as we stood up straight from our sitting position. [ZBR: 201–202] "At the instant that the actual *lift* takes place—not the whole process of standing up—are there any thoughts?" At this point, we did stand up from the sitting position. Many then reported (as I did) that no Self-centered thoughts had entered during *that* brief lifting moment. Next, when she invited us to stand up one more time, she then superimposed the word "What?"

At this point, the initial experience of *feeling less Self-referent* became even more noticeable.

She then went on to explain: "The loss of the sense of *I* occurs briefly when you move forward or bow into what is being done at this very *moment*. This loss happens by itself. It's not something you have to think about ... from now on, the approach in your practice is always to keep widening this gap of no-*I* ... so every time you bow or nod toward someone or something, you will be acknowledging, '*I* am *giving myself up*.'"

Our omnisense of Self is widely rooted. [ZBR: 193–200] It is also a very sticky construct. Like fly-paper, it's very difficult to become detached from. Perhaps Yamada Koun Roshi (1907–1989) was familiar with some of the same teachings about this selfless phenomenon during the act of standing up.[5] In his commentary on a koan, he said, "When you stand up, you simply stand up, there is only standing up in the whole universe, and the subject of standing up is emptiness."[6] The word "emptiness" being referred to here is interpretable at several different levels: Zen doctrinal, metaphorical, phenomenological, neurophysiological.[7] [ZB: 570–572]

The lower-level physiological mechanisms are visual, proprioceptive, kinetic, and vestibular. They can enter into many different kinds of movements. These influences become pertinent to some phenomena that occur during alternate states of consciousness. For example, researchers can blindfold subjects, then swing them in a large pendulum in different directions during a period of 2 to 20 minutes. [ZB: 724, n. 14] The subjects lose their sense of spatial reference, lose their sense of time, and go on to develop a wide variety of other alternate state phenomena. They may pay "visits" to other worlds and shift into religious and other mystical-type experiences.[8] People who experience vertigo discover how unsettling it is when their body-mind inhabits an unbalanced world.

Turning has more abstract meanings that enter into literary responses as well. Zen Master Bassui (1327–1387) noted that one particular meaningful word or phrase, called a *turning phrase*, can prompt a person to shift into a remarkable new dimension of seeing.[9] He was referring to the fact that many old Zen stories describe those persons who, after hearing only one word, have had their old concepts overturned and their eyes opened instantly to Buddhist knowledge during a major state of awakening. Robert Aitken Roshi, in a critique of one of Basho's *earlier haiku* about a crow on a withered branch, indicated that this crow *haiku* lacked maturity compared to Basho's different *haiku*, six years later, about the frog and the old pond. That later *haiku*, in contrast, could exemplify the sudden "turn of experience" at the instant the frog entered the water (chapter 24).

Turning Preferences in a Virtual Simulated Tunnel

A decade ago, Gramann et al.[10] studied the turning responses of subjects who navigated in a virtual reality tunnel that curved at the end. The researchers defined two types of navigation. *Path integration* meant velocity-based navigation. This type made internal *and* external judgments based on the subjects' impressions of velocity and acceleration. In contrast, *position-based* navigation used landmarks (chapter 6).

Their normal subjects fell into two groups. The so-called "turners" already started with a preferred egocentric strategy. They used it to make systematic Self-centered adjustments in their "homing" directions. Their "nonturners" did *not* have this preferred strategy.

The experiment offered optional approaches to both groups of subjects in the simulated tunnel. They were free to adopt an overhead, bird's-eye, allocentric perspective. Alternatively, they could take an egocentric perspective. When doing so, they imagined themselves riding a bike and *leaning into* the turns.

The original nonturners could be trained in the use of an allocentric frame of reference. Moreover, the original turners could adopt a fresh allocentric frame of reference without losing any accuracy. Accordingly, the researchers came to an important conclusion: "The use of an ego- or an allocentric frame of reference is *preferential* rather than obligatory in nature." This opens up another way to interpret some of the divergent results discussed in chapters 5, 6, and 9.

Intriguing Responses of the Brain When Normal People Turn to the Right

Baker and Holroyd followed up on their original finding that the *right* parahippocampal region was the origin of a curious effect on an event-related potential.[11, 12] In brief, they first focused on one particular topographical N 170 waveform that develops in a virtual T-maze environment. They found that this parahippocampal wave peaks 5 to 10 milliseconds *earlier* as people turn to their *right* rather than to their left for a reward. Moreover, during normal turnings in right or left directions, a *partial* phase resetting also occurs in the normal ongoing theta EEG rhythm. This rhythm normally cycles four to eight times per second. A partial phase resetting of this theta during turning means that its amplitude and phase coherence promptly undergo a brief increase. Both increased more during turning to the *right*.

Perhaps such differences seem negligible. However, suppose humans and other animals are given the option to turn *at random*. Now our normal spontaneous tendency is *to turn more to our left. Not* to our right.[13] Therefore, the actual rightward turning effect during these rewarded experiments is the kind of paradox that arrests researchers' attention. These changes in the N 170 waveform and in the phase of theta represent the brain *overcoming* its innate leftward turning bias. Some of the novelty involved in a shift that *turns* to the *right* is related to the presence in this task of a *context*-specific landmark. The asymmetrical physiological response appears almost to be subtly "highlighting" the location of the reward *if* the subject turns to the right, rather than to the left.[14] Such a covert signal is beckoning *subconsciously*.

The data also indicate that once a salient event or a local cue has caused this theta rhythm to be briefly reset, that person's short-term or longer-term potentiation more reliably encodes this new information into a memory that can be acted on. Notably, those subjects who do react with the larger N 170 potential can also draw more accurately from memory the spatial layout of a complex maze. This additional behavioral test confirmed that their earlier parahippocampal, event-related 5 to 10 millisecond change when turning to the right could predict their enhanced capacities to represent spatial relationships in general.

During fMRI monitoring, *two* regions in this *right* parahippocampal cortex become more activated by the rewards linked with turning right. These two right-sided clusters were located relatively far apart on this long gyrus. One cluster corresponds with the position of the anterior entorhinal cortex (figures 6.1 and 6.2). The other corresponds with a location associated with the posterior parahippocampal place area (PPA) (chapter 15). The legend of figure 6.2 indicates that farther forward, at least in the *hippocampus*, the proximal CA 1 nerve cells would be more strongly modulated by theta wave EEG activity (as is the MEC). The existing cross-talk between the LEC and the MEC renders it difficult to predict which net effect always predominates in the "anterior" entorhinal cortex, although it appears to receive more contributions from the LEC (chapter 6).

One factor in the psychophysiological bias that normally favors our turning left could begin over in our dominant left hemisphere. The motoric dominance of our left hemisphere is expressed in an obvious fact: most humans are right-handed and right-footed. Accordingly, most football and soccer kickers find that it is often easier to *lead* with their right foot. Once one's leading (more "dexterous") right foot has been placed out in advance of the left foot, this prior commitment ensures that it could now become easier to go on turning the rest of one's body toward the left. Try it.

It turns out that many of our leanings, turnings, and standings become forms of body language. Some body language can be translated into neurophysiological functions at levels in the brain that are relevant to Zen. Beyond that, the cultural origins of our "mind" language get very complicated[15] (appendix G).

13

Revisiting Kensho, *March 1982*

> *It is not that something different is seen, but that one sees differently.* It is as though the spatial act of seeing were changed by a new dimension.
>
> Carl Jung (1875–1961)[1]

It was 9 a.m. that March morning in 1982. Back then, I was in London, en route to the second morning of the two-day retreat. Suddenly, the spatial act of seeing entered an extraordinary new dimension. [ZB: 536–544] For more than three decades, I've been seeking to clarify which basic mechanisms could account for the huge shift into this new dimension.

As this millennium began, a new model suggested itself: *if* some triggering stimulus were to capture attention, a deep subcortical shift in the thalamus could transform the brain from its overconditioned Self-referential mode into an other-referential context.[2] But this problem remained: No single *big* stimulus had suddenly captured my attention that morning. My memory was very clear about that.

My incidental memory was also crystal clear about many other events during those few minutes leading up to 9 a.m.[3] For example, I did remember sitting by the window on the right side of the train as it traveled to this station. The scenery that had rushed past this train window on the right side could have contributed to some preliminary asymmetry of optico-kinetic stimulation in my two cerebral hemispheres. I also remembered descending from the train on that same right side, then turning slowly to the left and walking several steps.

And in recent years I began to search more carefully for links in the next sequences of remembered events. This inventory of items takes the following outline:

- Standing there on that unfamiliar platform, *watching* the train move away, *following* it as it curved to the right, became smaller, and then slowly disappeared down into that long tunnel

- *Hearing* the diminuendo of its clattering along the tracks, as this clackety-clack also followed that same route, and faded off into the distance

- Standing there now in the quiet of an empty platform, gazing at the dingy interior of this unfamiliar station

- Next, *turning casually, 45 degrees to the right, to look up,* beyond some grimy buildings

- *Seeing, way up there, a bit of open blue sky*

- Instantly, the *shifting* of this entire ordinary scene into a totally transformed novel dimension [ZB: 536–539; ZBR: 407–410, 414–432]

Tracking, Turning, and Looking Up

On the platform, both attentive acts—vision and audition—carefully followed the sight *and* clattering sound as the train gradually turned toward the right and disappeared from sight into the long underground tunnel. All these acts would coregister as a succession of dynamic changes in my three-dimensional visual *and* auditory perception.[4] Chapter 11 describes another person whose focus on auditory attention had a triggering effect as it tracked the sound of a jet airplane *moving* overhead.

Only this year would I learn how *turning to the right* preferentially activates the *right* parahippocampus in a virtual maze experiment[5] (chapter 12). Simply extending one's head back in order to look up at the clouds is an action against gravity, a movement that stimulates another part of the vestibular system. Turning and looking-up behaviors also stimulate the muscle stretch receptors on the back and along the sides of one's neck. The brain registers each such new change in position, aided by one's sense of proprioception.

So I wasn't "just" standing there inert. Multiple dynamic changes were converging in seemingly casual acts of sensory attention, perception, and behavior during these pivotal seconds on the train platform. When could a *sensitized* meditator's consciousness tip over into an alternative state of *kensho*? When the requisite configuration of such ordinary sensory changes happens to reach a critical mass. [ZB: 305–308, 346, 613–617]

Looking Up, Out There

Visual stimuli that enter our *superior* visual fields travel different visual pathways from those that enter *below* our visual horizon. Looking up, and *out there* at a distance, can correlate with potential shifts into allocentric processing. [MS: 20–37, 74–79, 88–91; ZBH: 110–119]

Recent evidence in human subjects suggests that the dorsal division of our reticular nucleus represents mainly this upper portion of both visual fields, whereas the ventral division represents the lower, opposite visual field.[6] [ZBR: 176–179; SI: 92–93]

The Novel Setting

I had not been in *this* train station before. What kinds of signals from a new scene switch the brain out of its usual rhythms and stimulate our sense of *novelty*? Kafkas and Montaldi found that the networks responsible for this novel activation are referable chiefly to our *ventral* visual stream.[7] First to be engaged are both the middle occipital areas (BA 18/19) and the fusiform gyri (BA 19/37). The next activations relay into the parahippocampus, perirhinal cortex, and hippocampus (chapter 6).

A novel scenic event does more than trigger a novelty signal in the hippocampus. It can also release acetylcholine elsewhere and prompt the nerve cells down in the substantia nigra and ventral tegmental area into a fresh release of their dopamine.[8] [ZB: 197–201] As several novelty mechanisms converge on the hippocampus and its connections, their "mismatch" responses confirm that an extraordinary new event has occurred, something their memory never registered before. [ZB: 219–220]

The affirmative qualities of insight-wisdom were extraordinary. Another feeling that March morning was also unique — a deep severing of all roots that had previously bound the psychic Self. Yes, this was a major, welcome, liberating sense of release (Skt. *moksha*). But later I would wonder: Why wasn't there something like the warm bliss that had infused the state of internal absorption seven years earlier?

The poetic literature of early Ch'an had anticipated this question. Its pages described the arctic visceral feeling in *kensho-satori* that could be associated with the dissolution of one's former Self. [ZBR: 436, 447] Centuries later, Master Yuan Wu Keqin (1063–1135) would be more explicit about what happens when the sharp sword of Prajna's wisdom cuts through all deep (limbic) obstructions. Then, he said, this "awesome chilling spirit" enables former entanglements to retreat by themselves, "without having to be pushed away."[9]

Having begun to link the activations and deactivations of different brain networks with the psychophysiology of different mental states, we continue in part IV to explore different correlations between other relevant clinical conditions and functional anatomical levels.

Part IV

Neurologizing

Only those contents of consciousness can be developed that correspond to the organization of the brain.

Walter R. Hess (1881–1973)

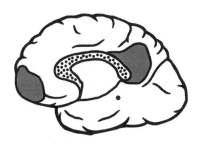

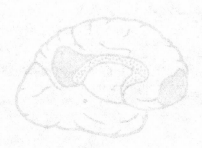

A Mondo in Clinical Neurology

If we knew what it was we were doing, it would not be called research, would it?

Albert Einstein (1879–1955)

The further I go, the better I see that it takes a great deal of work to succeed in rendering what I want to render: "instantaneity."

Claude Monet (1840–1926)

A *mondo* is an ancient form of Zen dialogue. [ZB: 110–111; ZBR: 64] Typically, when a monk is puzzled by one of the many paradoxes in Zen, we find him asking a reasonably good question, only to receive a tangential answer from his Zen master. The following pages do not shy from confronting a difficult problem in clinical neurology. We'll also be asking questions, and then trying to resolve potential answers in a rational manner. Notice that even though such a complex clinical problem does not yet yield one simple answer, it can still help us to appreciate some crucial normal functions of attentive processing. Ordinarily, these attentive sequences flash by so fast we never notice that they exist.

Our normal questions about *what* and *where* yield answers that merge seamlessly. It seems only natural to see both *where* some ordinary object stands out in *our* 3-D space and to recognize *what* it is instantly—a tall oak tree, for example. In contrast, certain neurological patients can *see* one such tree, but they can't see the whole surrounding forest at the *same* time. These patients are limited to a kind of one-tree-at-a-time, *piecemeal* perception.

Neural Mondo Question: Why *can't they see the rest of the forest*? This is equivalent to asking: Why can't they use their two categories of visual attentive processing—focal *and* global—at the same time? [ZB: 601–603; ZBR: 341]

Background for the Paradox

This disorder goes by the name, *simultanagnosia*. *Gnosis*, from the Greek, means to *know*. *Agnosia*, as coined by Freud, means a failure to know, a failure to recognize, a failure of meaning. In classical cases the patient's infarcts involve the upper posterior parietal—occipital cortex (*and* its underlying white matter) on both sides. Clearly, this *dorsal* posterior visual region normally helps process a wide variety of important functions. Moreover, its underlying white matter also conducts vital messages that normally pass to and from other separate regions that lie farther off in the distance, each having its own functions. Disconnections in this white matter, hidden from view, complicate matters enormously. Skilled electricians would be hard pressed to find a simple solution for a comparable problem. No wonder that neurological interpretations have come into conflict for more than a century.

At present, expert opinions still remain divided.[1] The situation is reminiscent of one of the Buddha's early classical teaching stories.[2] In this parable, the king invites each of several blind men to examine—by touch—separate parts of the same elephant. Each blind man has some tangible reason to be attached to his own contentious opinion. However, this interpretation is based on that one small part far removed from the representations of the elephant as a whole. In clinical neurology the situation is compounded: *different* patients with different pathologies are being examined by different neurologists who may cling to different belief systems.

The recent literature does not resolve all of these issues, yet some promising concepts are emerging. The following sample of reports does more than address our mondo question. It can also help clarify why meditators become aware during a retreat that their ongoing perceptions are now feeling so clear, instantaneous, and coherent. The reports suggest that these sensitized meditators could then be integrating more flexibly their normal, preconscious *focal and global* attentive processing skills out at the leading edge of their enhanced awareness. Some background information:

- Our *dorsal* attention stream normally uses its large, fast, conducting (M) cells to service our coarse-grained, low-spatial-frequency types of vision. Normally, we integrate these gross M-cell capacities seamlessly with those of the smaller, slower cells (P) in our *ventral* attention stream. These dual, parallel integrations proceed in a dynamic feed-forward and feedback manner. As a result, our *ventral* stream's finer-grained, high-spatial-frequency processing functions can proceed with even greater clarity, instantaneity, and coherence.[3,4,5]

- On the other hand, a crucial exception exists to such facile dorsal/ventral generalizations. It resides in some of the inferior temporal neurons down in the ventral stream. Some react *quickly*. In the monkey, when these sensitive cells fire in response to the *global* content of a visual display, their discharge occurs 28 milliseconds *before* other nearby cells fire to signal the *local* content of the display.[6] This heightened ventral sensitivity means that certain fast global cells are so tuned that they respond *quickly* to a large shape of low spatial frequency, almost as though they were M cells.

- Some evidence favors a conceptual model of simultanagnosia as a disorder that constricts a space-based (dorsal) system of attention. Such an unusual deficiency would restrict the patient to seeing only through a "narrow spatial window of attention."[7]

- In other patients with simultanagnosia, if researchers first make sufficiently salient the global form of their larger target, their patients can identify it.[8] However, their disorder lies in their *inflexibility* of top-down attention. This *inflexibility* of attention means that they can't select which stimulus is *relevant* to a given task.

- Moreover, this lack of flexibility interferes with more than their initial selection of the appropriate target stimulus. It also means that they can't shift *away* from this stimulus—can't *disengage* it and be *released* from it—after it has first been captured. [These last two examples, based on clinical findings in neurology patients, illustrate basic physiological principles. These principles are also applicable to meditators when they are deploying their normal skills of focused attention and/or global awareness (chapters 1 and 6).]

- In two other recent case reports of simultanagnosia, the patients' focus of preserved attention could still recognize target objects when these items were placed farther out in the visual periphery rather than just in the center.[9]

- The findings in one other recent case suggest that the disorder can express a physiological bias that favors *focal* attentional priority signals.[10] In this instance, the patient's parietal lesion appeared to have caused a "sticky fixation." This kept clinging to small focal targets. Clinging, letting go, and release from attachments are examples of words that have been in the lexicon of meditators for millennia.

Question: Could these substantial posterior parietal—occipital lesions (and/or their accompanying disconnections in white matter) *disinhibit* both colliculi far downstream in the midbrain? Could such lesions bias the colliculi to *over attend* to stimuli when they are presented at the central fixation point? These questions remind us that our superior collicu-

lus has remarkable visual and auditory reflexive functions.
[ZB: 240–242; SI: 80–89]

The Superior Colliculus

One superior colliculus is able to exert an early, potent influence on how we deploy attention to process a target that lies out in the visual *and* auditory field on that *same* side.[11] This "Sprague effect" was discovered in 1966. It reminds clinicians today that each superior colliculus is a key hub in the early mapping system we use to make instant orienting responses. The two superior colliculi stand poised to coordinate a covert, ascending, spatially adroit[12] "ancient visual grasp reflex."[13] These networks project up toward a succession of higher subcortical modules. Notably, this circuitry not only includes important nuclei farther up in our thalamus (pulvinar, intralaminar, medial dorsal, reticular). It also includes basal ganglia contributions,[14, 15] which help us select the most appropriate actions that orient to each stimulus.

In order for neuroscientists to untangle such long-standing semantic puzzles as simultanagnosia and "blindsight,"[16] we'll need to let go of certain upper-level preoccupations with the neocortex, pay more attention to the ways it interacts with some early, deeper regions that subserve the subconscious mechanisms of attentive processing (e.g., chapters 8 and 22), and tease out all the vanguard contributions of attention that service our focal and global processing.

Stay tuned …

15

Two Key Gyri, a Notable Sulcus, and the Wandering Cranial Nerve

Anatomy is destiny.

Sigmund Freud (1859–1939)

Neuroscience literature increasingly suggests a conceptual self composed of interacting neural regions, rather than independent local activations.

R. Murray and colleagues[1]

The first book in this series discussed those cortical and subcortical interactions most relevant to Self/other issues. [ZB: 146–290] Chapters in later books reflected the explosive growth of more refined neuroimaging techniques and experimental designs.

Readers today could still benefit from being reminded, as Murray et al. have just done, that we inherited a remarkably Self-centered organ, one that is also the source of our Self-inflicted longings, loathings, and delusions. And meditators, too, could benefit from having more information about how some simple neural Zen approaches retrain attentive processing skills that would otherwise lie dormant in their brains. [MS: 6–20] This chapter samples three regions of long-standing particular interest to creativity and to Zen: the Self, attention, and memory. We begin in the back of the inferior parietal lobule with these sobering reminders: each region performs multiple functions. Each enters into multiple, interactive networks.

The Angular Gyrus (BA 39)

As the concept of a so-called default network evolved, each paradox raised new questions. The subjects being scanned with PET and with fMRI were said to be "resting." Yet what explained all this high metabolic activity in their *medial* frontal and *medial* parietal cortex? And why would that small gyrus on the *outside* of the brain also share in this active "resting" configuration?

The angular gyrus occupies the posterior part of the inferior parietal lobule. The gyrus is larger in the left hemisphere (13.2 cm^3) than in the right hemisphere (11.7 cm^3). Its current cytoarchitectonic borders extend farther than some earlier indentations of the sulci might have suggested. A thicket of white-matter tracts converges on the angular gyrus and issues from it. They hint that it plays an active role in integrating the complex psychological functions of our heteromodal association cortex.

Suppose we framed this direct, neural mondo question: What does the angular gyrus "*do*"? Like a Zen master, an 18-page review article loses no time in undermining the premise of our question.[2] Seghier's answer: "The role of the angular gyrus cannot comprehensively be identified in isolation but needs to be understood in parallel with the influence from other regions."

Multiple reports confirm the high-priority status of the angular gyrus. It is a major higher-order cross-modal hub, not some isolated bump on the cortex. Meaningful (semantic) processing is the chief functional correlate when the angular gyrus is activated. This suggests that its activations, deactivations, and Self-centered functions become entangled during those conventional task-based fMRI experiments that use block designs. Moreover, it shares strong connections with the hippocampal and parahippocampal systems. These serve to caution any who would try to establish rigid divisions between ego- and allo-functions. Indeed, figure 3 in Seghier's review identifies 15 separate functions consistent with the angular gyrus as a whole.

Which basic mechanisms are involved in these 15 functional interactions? They seem more conceptual than perceptual. For example, the major processes themselves include these phrases: (1) semantic access; (2) the retrieval of facts; (3) the categorization of events; and (4) The shifting of attention toward relevant events.

How can one begin to further simplify the functional anatomy of the angular gyrus? By pointing to its three major subdivisions: (1) Its dorsal subdivision correlates with some "bottom-up" processing, (2) a ventral subdivision exerts a "top-down" influence during Self-referential processing, (3) a third, still more ventral subdivision participates in visual processing. It also joins in the other *deactivations* that take place in the *medial* frontoparietal regions of the so-called default network. Seghier's review concludes with this sobering reminder: "cracking the code that uniquely defines the angular gyrus function is still an ongoing endeavor."

With respect to its vital role in *meaning*, the latest studies all point to contributions from the angular gyrus during semantic processing in adults, normally developing children, and adolescents.[3,4,5]

The recent study by Douglas et al. from the National Institutes of Health is exemplary in its experimental design, execution, and results.[6] It dissects the key sensory/motor distinctions (in milliseconds) that separate our normal perceptions from our actions, including those willed acts that we make either subtly or deliberately. (Might it also suggest the potential for the angular gyrus to enhance some *motor* functions in meditators?) [ZB: 668–677]

- The researchers used event-related potentials [SI: 14–17] to target the delicate interval between that instant when a normal subject first notices the spontaneous *intention* to move the finger and the actual movement itself.

• The researchers also deployed high-definition transcranial direct stimulation (HD-tDCS). [ZBH: 164–166, 170] They showed that this benign low-amperage technique could modulate— up *or* down—those excitatory (+) or inhibitory (–) influences that three key modules normally exert on each other. Notice that the angular gyrus was one of these regions that entered into behavior. The two more obvious regions were the primary motor cortex and the cortex around the supplementary motor area.

What did the researchers find?

• When viewed as a simplified model system, the normal interactions in this triad could be represented as follows:

Stimulating the angular gyrus will enhance (+) the activation of the motor cortex.

Stimulating the motor cortex will reduce (–) the activation of the angular gyrus.

Stimulating the supplementary motor area will enhance (+) the activation of the motor cortex.

• Anodal tDCS stimulation of the angular gyrus enhanced the amplitude of the lateralized readiness potential from one to three *seconds before* the movement started.

• The results suggested that what we consider to be a brisk "movement" involves a long, drawn-out *subjective* process. Our first subconscious/conscious *intention* to move unfolds in parallel with, and overlaps, a second process—the actual *subconscious* motor initiation of the movement.

• The results support the efficacy of tDCS (an efficacy that some critics question.)

It suffices to note here that most transcranial stimulation methods are still too gross to reliably separate the angular gyrus per se (BA 39) from the supramarginal gyrus just in front of it (BA 40).[7] The supramarginal gyrus occupies the *anterior* part of the inferior parietal lobule. It has recently been parcellated into *five* structurally distinct regions. It is clearly no longer appropriate to speak about the "inferior parietal lobule" as having only one function.[8]

A different gyrus occupies the medial temporal lobe. Its multiple functions are outstripping the impulses of histologists to split it up into several subregions with different names.

The Parahippocampal Gyrus

We rarely see a picture showing the *under*surface of the temporal lobe. This is unfortunate, considering how crucially important it is for the *ventral visual processing stream*. Figure 15.1 helps redress this imbalance.

It shows the long parahippocampal gyrus as the innermost (most medial) gyrus. Posteriorly, it blends in with the lingual gyrus. What can't be seen from the outside is the way the parahippocampus blends into the perirhinal cortex in front[9] and this region then curls up to become what is called the hippocampal formation (figures 6.1 and 6.2).

The collateral sulcus is the lateral boundary along the lateral edge of the parahippocampus. This sulcus is a valley that serves to separate the parahippocampus from the adjacent fusiform gyrus. [SI: 19, 25, 132; MS: 25, 132; ZBH: 124, 131] The long inferior temporal gyrus lies way over on the outside (most lateral edge) of each temporal lobe.

A recent map of the human cortex defines the 25 different regions that we use to process our refined sense of vision.[10] Chapter 6 introduced the parahippocampal gyrus as a key contributor to this map. Our primary visual areas, V1 to V3,

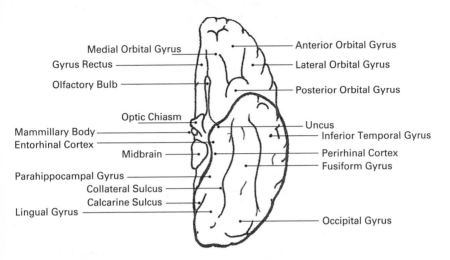

Figure 15.1
A basal view of the left hemisphere, emphasizing the temporal lobe.
The frontal lobe is at the top. The occipital lobe is at the bottom. Notice that the *parahippocampal gyrus* leads (imperceptibly) first into the *perirhinal cortex* and then into the *entorhinal cortex* as it nears the (hidden) hippocampus. Proceeding from the midline at the left and moving laterally toward the right, the three largest temporal gyri begin with this *parahippocampal gyrus*. Next in line is the *fusiform gyrus*, and finally the *inferior temporal gyrus*. The *uncus* is a major entry site for the uncinate tract. Its two-way connections link our temporal lobe functions with those of the inferior frontal lobe. In the hypothalamus, the *mammillary body* issues the important tract that connects the hippocampus with the anterior thalamus. The nomenclature follows that in J. Mai, J. Assheuer, and G. Paxinos. *Atlas of the Human Brain,* 2nd ed. (San Diego, CA: Elsevier, 2004), 92, 121–219.

are first in the visual hierarchy. These begin far back in the occipital lobe. Next in succession is that long fusiform gyrus (50 mm) with its visual association areas, termed VO1 and VO2. Then, beyond these, are the two parahippocampal visual areas labeled PHC1 and PHC2. What is so special about these two visual regions of the PHC?

- They respond strongly to the spatial aspects of *scenes*.

- Their *upper* visual field representations share a foveal representation *centered* next to the midline. Another related fact was discovered in monkeys over four decades ago [chapter 5].

- These regions are strongly biased toward representations of the visual fields that are more peripheral (*global*) in nature.*

- Some of their responses can overlap those associated with parahippocampal regions that in the past have been regarded as representing parahippocampal *place* areas (PPA) (as discussed earlier in chapter 6).

Weiner and Grill-Spector[11] have demonstrated not one but *two* distinct *face*-selective regions, 15 mm apart, on the fusiform gyrus. These *fusiform face* regions (FFA) are part of a recurring pattern of regional clusters. As impulses flow through our parahippocampal gyrus, these separable configurations are coded to relay crucial information about *other* persons' faces *and* limbs into the relevant environmental scenery of the moment. How can all this help a human social animal? Consider how fortunate we are to be able to

---

* These dual capacities are conferred by different nerve cells. Their merger could enable PHC 1 and PHC 2 to blend some assets analogous to those of a wide-angle lens with others resembling those of a convex lens. Under conditions of rapid parallel processing and ventral impulse flow, the subjective experience during the state of internal absorption can be remarkable (chapter 16). [ZB: 482–485]

perceive (and then "read") the facial expressions and the "body language" of other persons.

Weiner and Grill-Spector hypothesize that visual percepts become transformed into the codes we use for our long-term memory during these steps that relay impulses starting in the earliest visual areas all along this ventral occipital → temporal pathway. Their model also suggests that some pathways that begin *laterally* develop special processing functions as they move medially. For example:

- Relays from the *lateral* occipital-temporal sulcus that lead into the medial fusiform gyrus enable us to recognize form, objects, and faces. Then, from this fusiform region on,

- Transmissions pursuing a more *medial direction* converge into the still more medial-lying whole parahippocampus gyrus. These enable visual impulses to be represented and registered as percepts as they continue their course along the lateral and medial pathways of the entorhinal cortex (LEC, MEC). (Figure 6.2 illustrates the potential for cross-talk to occur between the LEC and the MEC.)

Bastin and colleagues placed *intracerebral* recording electrodes within the posterior parahippocampal region in nine patients.[12] They focused on responses consistent with those of a parahippocampal "place area" (PPA). They asked: Why was this parahippocampal "place" region sensitive to something *more* than just a simple visual scene? Why did it also respond more strongly to larger *objects* (objects that could stand out and were therefore *more* suitable as landmarks) than to smaller objects that were smaller, closer, and took up less space (therefore were items *less* suitable as landmarks)?

They showed pictures in gray scale to their subjects. The images in one experiment represented different categories (e.g., separate scenes, faces, animals, objects). Then, in a second experiment, they showed full-color images of scenes in different categories. A cluster of posterior sites gave rise to the subjects' strongest gamma EEG responses (at 50–150 cps) to these colored scenes. These sites were in and around the depths of the *collateral sulcus*. This posterior cluster lay about 10 mm *behind* the boundary separating the parahippocampal gyrus from the lingual gyrus. [ZBH: 131]

The subjects' earliest responses, at 80 milliseconds, served to discriminate the presentations of scenic stimuli from those of nonscenic stimuli. This short interval suggested that a bottom-up process was encoding scenes that contained specific visual or geometric features. Not until around 170 milliseconds did the gamma response serve to differentiate buildings from non-buildings. This later processing peak suggested that other spatial or semantic information coding for features could have been arriving from distant cortical regions. Chapter 6 raised the possibility that *if* these later messages were of navigational relevance, they might have arisen from the retrosplenial cortex and precuneus higher up in the medial parietal cortex.

The parahippocampus also enters into reciprocal connections with the frontal lobe that can escape notice.[13] This "parahippocampal-prefrontal system" is on the list of outflow pathways from the medial temporal lobe (chapter 6). How this anterior pathway contributes to our complex sense of familiarity, retrieval, and other cognitive functions is still under investigation.

The Superior Temporal Sulcus

In a book mostly concerned with "peak experiences" and bulging gyri, why bother with our brain's less conspicuous valleys? The superior temporal sulcus (STS) is one of the longest and most cognitively fertile valleys in the human brain (BA 21, 22). It establishes the lengthy boundary between the superior and middle temporal gyri. Earlier reviews sampled its contributions to our sense of meaning, figure-ground relationships, humor, and to loving-kindness meditation. [ZBR:152–157; SI: 24, 41, 137, 142, 146; ZBH: 205 n. 6]

Recently, Deen et al.[14] used fMRI to monitor the ways 20 adults responded during five separate tasks: (1) theory of mind; (2) biological motion; (3) dynamic face perception; (4) voice perception; and (5) auditory story plot. In brief, this sulcus was found to contain multiple relatively domain-specific regions as well as regions that responded to multiple types of socially vital interpersonal information.

In another fMRI study,[15] the right posterior STS became activated early and independently as soon as the subjects interpreted different feature-based facial expressions. In contrast, the right inferior frontal gyrus showed a more holistic pattern of responses to these same facial expressions.

How do we first analyze the situation when we hear the other person's voice rising in anger? Korb et al.[16] monitored their subjects' responses with fMRI. Their initial processing of this voice of wrath involved the *right* superior temporal gyrus, the superior temporal sulcus, and the inferior frontal gyrus.

Human responses to human *voices* (not just to mere sounds) preferentially activate the superior temporal sulcus and gyrus (chapter 10). Three clusters represent these voice-sensitive regions. They lie within the posterior, middle, and anterior regions of the superior temporal sulcus and its borders with the superior temporal gyrus. A fourth group of less voice-sensitive areas occurs in the inferior prefrontal cortex and amygdala on both sides.[17]

The Wandering Cranial Nerve

The vagus is our tenth and longest cranial nerve. [SI: 173–176, 225–227] The two vagi play vital sensori-motor roles in our vocalization, swallowing, cardio-pulmonary reflexes, gastrointestinal and immunological functions.[18] Moreover, unilateral left vagus nerve stimulation has been used to help patients who are epileptic or depressed. This suggests that some vagal projections *ascend* to influence key *cerebral* aspects of consciousness, directly or indirectly.[19] But how?

Certain programs of meditative training, including an eight-week MBSR course, are also reported to benefit not only cognitive/affective functions but also to improve the rates at which patients recover from severe, recurrent depressions (chapter 1). Again, how does *this* benefit occur?

So, could vagal nerve stimulation, and meditative training, share certain mechanisms in common that enables them to benefit such patients? For example, might they each help sponsor a more "optimal balance" among the complex relationships that exist in the "classical" *central ascending* vagal projections?[20] One of these [less expected] upper level components is the central norepinephrine projection system. [ZB: 197–208] This synaptic cascade proceeds chiefly from the nucleus of the solitary tract up to the locus coeruleus. [ZB: 165] The locus coeruleus then releases norepinephrine into multiple sites on up to the cerebral cortex.[21] Why are such excitatory and inhibitory steps so difficult to isolate and identify? Because these many networks' interactive connectivities and receptor-based responses are integrated so intimately into our basic arousal, awareness, attention, and recognition memory functions.

The Zen Way values direct experiences and body language much more than wordy concepts, endless discriminations and abstractions. [SI: 217–218] The next chapter illustrates how directly experienced, first-person visual phenomena can also provide clues that help to interpret meditative states of consciousness.

16

Paradox: The Maple Leaf Way Up in Ambient Space

Just an old leaf,
yet try to follow its structure—
or count its colors!

J. Hackett[1]

One evening back in December 1974, while meditating in a zendo where I'd never been before, I dropped into an extraordinary state of internal absorption. [ZB: 467–473; ZBR: 315–322] At the start of this enhanced phase of awareness, a vividly colored red leaf materialized in my vision. *Where* was this leaf? Way *up* in the *top left corner* of this vast black, silent space.

Who was witnessing it anonymously? A totally aware nonentity. *What* was so peculiar about this leaf? It was way too far up beyond my brow to be clearly focused on by any brain constrained by its normal field of vision. Even so, this leaf's fine-grained detailed structure was still being *seen as clearly* as though it was centered *directly in front of the viewer*. [ZB: 482–487] Yet that's impossible!

On the other hand, chapters 5, 6, and 15 have kept reminding us: nerve cells in the normal parahippocampus and inferior temporal cortex have special visual attributes. When their properties are enhanced, these neurons could blend focal functions with overlapping global functions in this seemingly paradoxical manner.[2]

But back in 1974, I was surprised when my 35-mm transparency slides arrived in the mail two months later. They proved that this photographer had forgotten an important fact. They reminded me that I had previously taken several pictures of this very same Japanese maple leaf. Indeed, I had focused more attentively on this one real leaf than on any other single item in recent months. *When*? Seven weeks *before* that evening in the new zendo. Seven weeks *before* this same red leaf had then suddenly reappeared, this time as a formed hallucination way up in the space to my left. Only in retrospect did the implications of this striking visual hallucination sink in.

The Lapse of Memory

What had happened to that earlier memory—the memory that had first registered *my* acts of clicking the shutter on that red leaf? It had been relegated to some separate, working memory compartment, stored there, then forgotten (as most working memories are). Events laid down during one state do tend to become "state-bound." However, in that unfamiliar zendo seven weeks later, the full stream of conscious awareness during internal absorption seemed to have been flowing much faster and deeper. During this dynamic state, the rapid rate of highly attentive processing, now surging at full volume down a globally widened mainstream, had swept up that forgotten maple leaf on the bank and projected it into a single, vivid, full-blown hallucination. [ZB: 461–462]

In the decades since, we've been learning much more about how memories are normally laid down and later recollected (as discussed in chapters 7 through 10). We've also seen how focal and global aspects of perception can overlap their functions physiologically. The next chapter reminds us that buried traces of our memories are never far removed from being under the influence of *nitric oxide*. In 1998, this signaling molecule would be another topic that merited Nobel Prize–winning recognition.

The Nitric Oxide Connection

During the last decades, nitric oxide (NO•) has emerged as a critical physiological signaling molecule in mammalian tissues, notably in the brain.

R. Santos and colleagues[1]

In the pursuit of knowledge, every day something is added.
In the practice of the Tao, every day something is dropped.

Tao Te Ching

The recent review of nitric oxide (NO•) by Santos et al. is recommended. Its 162 references illustrate why this gaseous signaling molecule continues to play so many diverse, relevant *yin/yang* roles in the brain. [ZB: 412–413, 655–656; ZBR: 279–288; SI: 260–261; MS: 138, 207 n. 2; ZBH: 196] This chapter focuses on recent articles relevant to those multiple aspects of our memory functions already discussed in chapter 6 and subsequently.

The Synthesizing Systems That Generate NO•

- Nerve cells in the normal brain contain two different enzymes that synthesize NO•.[2] The first of these *neuronal* enzymes is widespread (nitric oxide synthase 1). The neuronal steps begin with the release of glutamate. This prompts the enzyme in nerve cells to synthesize NO•.[3] The second neuronal enzyme is called nitric oxide synthase 2. It is expressed only transiently in nerve cells of the piriform and entorhinal cortex, dentate gyrus, medial thalamus, hypothalamus, and cerebellum.

- However, the lining of smaller blood vessels contains a third enzyme that synthesizes NO•. This *endothelial* enzyme is called nitric oxide synthase 3.[4] As soon as it releases NO• from the inner lining of blood vessels in the brain, vasodilation mechanisms help match the requisite increased flow of oxygenated cerebral blood with the metabolic requirements of the firing nerve cells.

NO• in the Hippocampus

- The diffusion of NO• also produces major effects on the activity of other nerve cells. These mechanisms are separate from those cited above that increase their supply of blood.

- In the hippocampus, impulses flow normally from the dentate gyrus to the CA 3 subregion and then on to the CA 1 pyramidal cells (figure 6.1). NO• is promptly released locally whenever glutamate is added to each site separately. Slightly less NO• is released by the dentate gyrus, slightly more by CA 3. The slightly slower release of NO• by CA 1 cells is more prolonged.[5]

- The normal maze-learning and memory behaviors of rats can be enhanced by increasing the levels of NO• or by increasing the activity of the nicotinic acetylcholine receptors. Moreover, these two systems act synergistically.[6] In contrast, reducing the levels of NO•, increases the inhibitory effects of GABA A receptors in mouse CA 1 hippocampal slices.[7]

- Spatial memory tests show that inhibiting the nitric oxide synthase activities in adult rats interferes with their memory acquisition, consolidation, and retrieval.[8]

- When rats react to the acute stress of being restrained, they enter into a state of major behavioral and autonomic excitement.[9] This is associated with a substantial increase in the activity of their normal glutamate → NMDA receptor → neuronal nitric oxide synthase → NO• → cyclic GMP pathway. It leads to a further cascade of delayed second messenger metabolic changes. [ZB: 235–240] Delayed changes can provide clues to the nature of earlier events (chapter 9).

Comment

For these reasons it would seem a worthwhile empirical step to use diffusion-weighted imaging techniques to study those few meditators that have just undergone a major acute spontaneous change in their states of consciousness (appendix D). Why? Because during the major states of *kensho-satori* or of internal absorption, it is plausible to hypothesize that substantial dynamic changes could have occurred in relevant messenger systems. Because NO• can be released from numerous neuronal and endothelial sites, one awaits the different empirical results with great interest.

The transformations of consciousness during *kensho and satori* decondition the brain of prior maladaptive traits. [ZB: 327–224] These phenomena of liberation have been known for millennia. [SI: 190–193] It feels like "the bottom of a water bucket has just given way." In Buddhism, such a release that cuts the old psychic bonds is called *moksha*. [SI: 207–211]

In this regard, nitric oxide could serve as a two-edged sword, a collaborator in the mechanisms that can also cause a *delayed* death of nerve cells. [ZB: 654–659] For example, the *overfiring* of glutamate and aspartate nerve cells can render both CA 1 and CA 3 nerve cells nerve cells vulnerable. Excessive release of these excitatory amino acid transmitters does more than co-activate the synthase that makes more neuronal NO•. It also activates another enzyme, calpain.[10] The activation of calpain unleashes a cascade of cell death mechanisms (*apoptosis*). As nerve cells die, their lysosomal digestive enzymes are also released, a process that can lead to cavities.[11] Diffusion-weighted imaging (DWI) can document similar local, delayed changes in the brain, as has been shown to occur in transient global amnesia (chapter 9). Learning new ways to perceive, interpret, and react means pruning old maladaptive dendrites, cell bodies, axons, and synapses.

Curiosity serves a time-honored role in biomedical research.[12] At this writing, fMRI and DWI studies have not yet been performed either in acute states of *kensho-satori* or in acute states of classical, major internal absorption. This now seems an appropriate time to search—simply on an *empirical* basis—for potential sites where neuromessengers have been released in ways that are now known also to release NO• or other toxic molecules. Such acute states have delayed effects on the brain, both constructive and destructive. These changes could still be discernable using diffusion-weighted imaging, when such studies are performed not only in the early hours but also three to five days *after* their onset.[13]

To the degree that basal nitric oxide levels may be higher in certain normal subjects than in others, serial determinations of nitric oxide in blood and spinal fluid could also be informative. For example, NO• levels in blood are said to be 55 percent higher in a group of migraine patients than in their controls.[14]

Young mice can inherit a partial genetic deficiency of endothelial NO• synthase. They develop premature thrombotic occlusions in the cortex of their temporoparietal, retrosplenial and hippocampal formation.[15] Figure 7C in that report illustrates that their amyloid angiopathy involves the lateral aspect of their dorsal hippocampus.

In the next chapter, we turn to other neuronal mechanisms that cause visual items to become salient.

18

"Pop-Out"

> A red cherry among green leaves seems to pop-out effortlessly
> in our visual experience.
>
> T. Ossandon and colleagues[1]

In ordinary daily life experience, it's relatively easy to distinguish the two levels of conscious and subconscious processing. Obviously, we can focus our attention deliberately, voluntarily, and become fully conscious of what we perceive. On most other occasions, pre-attentive processing operates automatically, preconsciously, and we are not made fully aware of each one of its covert operations. [ZB: 278–281; MS: 147–150]

Zen practices have always involved (1) the training of our attention, (2) the dissolution of our maladaptive Self-centeredness, and (3) the retraining of an other-oriented lowercase self. The pop-out phenomenon offers one of several ways to measure how our attentive and perceptive skills could evolve during this long-term meditative approach. [SI: 3–48]

The term "pop-out" usually describes what happens in a visual experiment. Even though the researchers present you with a visual field full of distractions, one special item captures your attention. Like a red cherry set against a green background, this item leaps forth instantly into consciousness. [SI: 35–36]

Hayakawa and colleagues[2] monitored 10 subjects with magnetoencephalography (MEG) [appendix C]. One oblique bar was their target. It was hidden among the 35 other vertical bars that served as distractors. As this single oblique bar popped out, one waveform in the subjects' MEG became much larger. This peak arrived at 196 milliseconds. It appeared simultaneously in each of *two* activated regions: (1) the intraparietal sulcus (IPS) and (2) the posterior part of the superior temporal sulcus (STS), R > L (chapter 15).

The occipital lobe showed two peaks in its calcarine sulcus. The early peak, at 110 milliseconds, was consistent with the way our primary visual cortex initiates the processing of elementary visual features. The later peak, at 250 milliseconds, identified a recurrent *feedback* process. This plausibly reflected the *earlier* steps in attentive processing, namely those impulses that had been relayed back from the two, earliest, *parietal and temporal* sites described in the paragraph above. The parietal site is a recognized part of the *dorsal* attention system. The temporal site lies within a larger zone (the "TPJ") described as part of the *ventral* attention system. [SI: 27–34] The evidence in this MEG experiment suggests that "pop-out" could enlist *both* dorsal and ventral attention systems *simultaneously*. The greater activity in the *right* STS suggests that pop-out was activating more of this reflexive, "circuit-breaking" attention system on the right side. [MS: 101–102; ZBH: 201 n. 3]

Recent research using intracranial EEG electrodes[3] supports this pioneer study. The 24 patients were *deliberately searching* for one target. This letter **T** popped out from among the 35 letter **L** distractors. During this particular *voluntary* search, gamma-band activities increased throughout the whole *dorsal* attention network. The investigators noted that any such deliberate, sustained, "active," serial visual search will necessarily activate these dorsal, *top-down*, parietal → frontal attention networks. This would not exclude the possibility that coexisting parallel reflexive influences could also be arising more globally from the "bottom up."

Whether these patients' prior search efforts had been efficient or inefficient, most EEG responses showed similar onset latencies and amplitudes. Yet when their searches were *more* efficient, certain other transient responses arrived "relatively late" in their occipital and temporal cortex (at around 250 to 400 milliseconds). Again, these late responses were consistent with the delayed arrival of feedback, representing impulses that had been further refined during the earlier visual processing steps.

Letting It Happen

For years we've heard that it's often better to be openly alert and "*let* it happen" than to try too hard to *make* it happen. Watson et al.[4] emphasized an important point: performance can be substantially enhanced when we *allow* a relatively *passive* cognitive strategy to guide our attention more or less *involuntarily*. Their behavioral study is especially relevant for meditators who need to learn how to *let go* in order to enter into the more passive, *yet still dynamic*, awareness that characterizes receptive meditation techniques (chapter 1). [MS: 42–52]

This study divided the 44 university students into two groups. The *passive* group's instructions were: "Be as receptive as possible and let the unique item 'pop' into your mind as you look into the screen." In contrast, the *active* group's instructions emphasized a deliberately active effortful search. All targets were circles that contained only a *single* gap in their circumference. This gap was located either on its right *or* left side. The distractors were also circles. However, these distractors all had two gaps, one on *each* side. An eye-tracker carefully monitored how the subjects moved their eyes. They pressed on a button to signal when "pop-out" occurred.

The results proved interesting:

- Those students who were instructed to become passively receptive had faster reaction times ($p < 0.01$). Yet this increased speed coincided with more errors.

- The subjects using this passive approach also searched more efficiently ($p < 0.05$). Which measurement was the best index of this search efficiency? The best test that separated these two groups was the latency between the saccade to the target and the corresponding button-pressing response. (This interval usually took between 400 and 800 milliseconds.)

- The passively instructed students made only three or fewer saccades before they fixed on the target ($p < 0.001$). In contrast, the actively instructed students kept making more saccades, even after they had already fixed on the target.

- The successful searchers in the passively instructed group also made larger-amplitude saccades to their targets.

Sensitive Areas versus "Clinging" Areas

Certain areas of the posterior cortex are known to be most sensitive to a particular class of visual objects. Important among such dedicated areas are the face area in the fusiform gyrus (FFA) and the place areas in the parahippocampal gyrus (PHPA) (chapters 6 and 15). It is no surprise that these same dedicated regions tend to be *less* influenced by any clutter of surrounding distractions in the visual scene.[5] On the other hand, recent fMRI studies of pop-out have suggested a kind of "clinging" role for activity that localizes to the *right* inferior to middle frontal gyrus (IFG). This localization occurred at those particular times when the subjects could *not* easily *disengage* their attention from the presence of the identified target stimulus[6] (chapter 14). This finding supports the earlier evidence that these two right-sided frontal gyri play an "executive" role in normally helping to integrate the functions of our dorsal and ventral attention systems. [SI: 31]

Of course, all the tasks described above took place in the ivory-tower setting of a research laboratory. How would *you* like to *play* outdoors, with your brain so accustomed to seeing a small white object speeding toward you at 139 feet per second that it automatically adjusts your whole body to it, with no effort on your part?

19

Keeping Your Eye on the Ball

> Tell me to what you pay attention and I will tell you who you are.
>
> Jose Ortega y Gasset (1883–1955)

> I was never in control of the zone. Rather, it passed through me as it pleased.
>
> Shaun Green[1]

Training attention is the fundamental meditation skill. Suppose you've already started a regular, formal meditative practice. Let's hope that you will have been cultivating more than an awareness of your body-mind in the present moment, more than mindful kinds of introspection, more than the intuitive recollections of remindfulness. Perhaps now that you've become increasingly competent in these three basic skills you will also have noticed something else: maybe you've also been initiated into the artless art of *letting go*.

Letting go—*where*? Into the flow. Letting go—into *what*? Into the most efficient forms of *effortless* behavior (Ch: *wu-wei*). [ZB: 668–677; ZBR: 322, 361; MS: 37–41, 47–49] Back in feudal Japan, some samurai warriors would learn, from direct first-hand experience, that Zen training enhanced their physical and mental survival skills. In today's frenzied professional sports, great baseball athletes like Shawn Green and legendary basketball player-coaches like Phil Jackson[2] describe how a meditative, attentive approach toward "playing in the zone" enhanced both individual and team performance.

About Shawn Green

Green first read about Zen when he was a senior in high school. Back then, he practiced *qigong*. Later, on the batting tee, while practicing his swing on the incoming ball, he finally started "taking each swing at full attention." Before long, he realized that "being fully attentive is being fully alive." Ultimately, "I became the act of hitting rather than the person who was hitting." No overinflated Self was clinging to that act of hitting. Instead, he was tapping into subconscious *procedural* memories. [MS: 134, 138]

Fast forward to May 2002. Green is now an outfielder for the Los Angeles Dodgers, awarded with All-Star status. Pressure on him intensifies because he is also recognized as the most outstanding Jewish player since Sandy Koufax and Hank Greenberg. But now there's one huge problem: during these past six weeks, Green has been in a slump. He's hit only three homers and often gone hitless. Will he take a break—goof off? No. He decides to hone his attentive skills during more batting practice. Finally—after this *extra* practice at hitting a ball—those swings finally *become effortless*. He *lets go* into the *act* of hitting. Hitting *happens*. He isn't trying so hard to make it happen. He's tapping into even deeper, *selfless procedural* memories. [MS: 134, 138]

It's now the ninth inning of May 23, 2002, late in the afternoon. The pitch is a fast ball. That small white object speeds toward him going 95 miles an hour. Even so, it seems to arrive in slow motion. With a sharp "crack," bat meets ball, sends it sailing over the fence in right-center field. Home run!

But, notice what else had happened that afternoon. Green had made five earlier trips to the plate. By then, he'd already hit *three* home runs, one double, and one single. So, this latest homer was his *fourth*. He has just set the all-time Major League hitting record: *19* bases in one game! Four of them home runs!

Green keeps a journal. His comment for that day was the essence of procedural learning. It reads: "Everything was working now on its own." During the next five games, the momentum of this earlier "on-*its*-own" streak allowed him to hit *nine* more homers.

Commentary

When a person becomes 100 percent "in the zone," the *powers of attention focus spontaneously*. A recent book, edited by Brian Bruya, devotes 16 chapters to this topic of *effortless attention*.[3] The chapter by Wulf and Lewthwaite is important for meditators and sports enthusiasts alike. Why? Because they found that when athletes specifically focus attention on an *external* target they increase the efficiency, accuracy, and effectiveness of their movements. Moreover, this *external* focusing also *decreased the associated* mental effort. This result was a sharp contrast with those athletes who had chosen to focus internally either on themselves or on parts of their bodies. This *other-focusing advantage* showed up in the results using many different tests for performance: accurate free-throws in basketball, soccer kicks, jump-and-reach heights, and treadmill running. The results are also documented by electromyography (EMG) during weight lifting. Could it become counterproductive for an individual meditator to focus excessively on scanning his or her own body?[4]

By way of anecdotal asides, I rediscovered one similar external attention-focusing technique a decade ago. It still helps me serve more consistently in tennis. The technique was simple: just focus attention precisely. *Where?* On some detail of the ball. *When?* Especially at the apex of each toss.

Clearly, this was only another reminder of the same old platitude I had heard about and practiced back when I was a kid: "Keep your eye on the ball!"

Now, at 91, it's obvious to me, and to other observers, that my mind/body connection also works best when I stay fit by exercising regularly. [MS: 118–119] When do I become much clearer during the present moment, more efficient physically, more remindful of events in the past, and sleep better in general? The day *after* I've worked up a sweat for half an hour in the gym twice a week, played doubles tennis once a week, while maintaining the regular morning routine of setting-up and yoga exercises for 15 minutes, followed by 25+ minutes of open-eyed meditation. Confirming this impression is the report that older adults who maintain higher levels of aerobic fitness turn out to have larger hippocampal volumes and better spatial memories.[5]

Part V

Living Zen

Zen is not about being irresponsible and ignoring things; it is about being able to remain unattached, about seeing clearly, and beginning again, and letting go of the mind that hungers on something over and over again.

Shodo Harada-Roshi

What Is Living Zen?

When waking up, really pay attention. But make no effort to struggle with what may be going on in your mind. Struggling wastes energy. One gains energy by preserving it right in the midst of the mundane stress of daily affairs. It is here that you obtain buddhahood. Here, you can turn hell into heaven.

Ch'an Master Dahui Zonggao (1089–1163)[1]

Don't overemphasize those dramatic, brief, states of awakening called *kensho* or *satori* in Zen. Authentic Zen training places a major ongoing emphasis on how one's attentive competence on the cushion becomes *incrementally* transformed into an actualized, "Living Zen."

Living Zen is that much longer, much slower development on the Path. It refers to one's *direct, lived experience out in the real world*, leaning into just one activity at a time—attentive to and aware during each present moment. It means that one's posture and attentiveness are totally engaging *this* moment—*just this* moment—whether the act is eating, reading, listening to someone else, or looking up casually at cloud formations high off in the blue sky. [ZB: 76–77]

Gradually, one's former mind*less*, inattentive behaviors start registering more mind*fully* in awareness. Those old, habitual maladaptive ruminations become more positive, efficient, *re*mindful and adaptive. Intuitions start ripening into insights on one's way toward simply becoming wiser and more genuinely compassionate toward oneself and other persons. Living Zen slowly realizes these developments more selflessly over a time frame of years and decades, not just a few weeks. Can such a long-term meditative approach actually help dissolve the pejorative Self? [ZB: 141–145]

It can, only to the degree that a Living Zen moves beyond what one learns on the cushion and then integrates similar practices into one's everyday life (J. *shugyo*). *An earthy, flexible empiricism resides in the core of Living Zen.* [SI: 13, 203–205] Yes, it does include becoming more in touch with each new event. Yet this also means welcoming it in a simpler, matter-of-fact manner. Indeed, a new level of open awareness will gradually shed any former possessive, Self-conscious attachments to earlier levels of more explicit perception. Finally, with fewer possessions clinging to that former Self, the "zone" now *starts passing through one as it pleases* (chapter 19).

Exercised by the challenging encounters in daily life, the practices of a Living Zen slowly cultivate one's innate attributes of clarity, wisdom, compassion, rectitude, and altruism (chapter 3). Simultaneously, the practices also help develop enough savvy to steer clear of situations that invite such unfruitful behaviors as Self-indulgence, spiritual materialism, and the use of intoxicants.

Living Zen develops a profound sense of gratitude to all others who have helped in the past. This ongoing gratitude practice accesses deep levels of remindfulness. [MS: 140–145] Gratitude has recently been induced and studied in normal subjects after they first imagine themselves to be Holocaust survivors.[2] The resulting emotional fMRI responses can be correlated with enhanced BOLD signals. Two regions are chiefly involved: (1) the anterior cingulate cortex around the genu and (2) the medial prefrontal cortex in its ventral and dorsal divisions. [SI: 25, 209–211]

Living Zen is grounded in retraining and repetition. One repeats fruitful behaviors until they become habitual. Only these *repeated* wholesome daily life practices will become the agency of the incremental brain change that helps develop character. This process is now referred to using the magical word *neuroplasticity*. In decades past, it was just called "learning from experience."

During moments of solitude, Living Zen also learns how to shift more spontaneously into an introspective phase. In quiet settings, small insights and reappraisals can emerge from *deeper incubations*. They still go on—working day and night—at covert, wordless, subconscious levels. [ZBR: 31]

Living Zen is not easy. It requires the utmost hard-won sophistication of a subtle, silent, emerging remindfulness. *It means making repeated selfless, skillful, adaptive behavioral concessions to everyone else with whom you interact*—in your family, at your job, in your *sangha*. It means subordinating your pride, with all *its* urgent precious "needs," to the real hard facts and issues at hand, adapting in new ways that create more effective harmonious relationships. Living Zen inhabits the increasing scope of this more objective spaciousness, exercising new-found degrees of courage, *clarity*, openness, gratitude, and lightness of behavior.

The unannounced goal? A simpler, more humane being. A person whose increasingly subtle capacities for generosity and compassionate behavior now extend outward so skillfully that they include not just the rest of humanity and the whole environment at large. They also extend inward, nourishing the no less legitimate interests of a smaller-sized, lower-case self. How might you sense the presence of this formless, more flexible self? Explore carefully, after you've been meditating repeatedly on a retreat. You'll discover its outline there, inside the more spacious zone, where your former rigid, suffering and insufferable, *I-Me-Mine* center used to be. [MS: 137–139]

Sometimes, Zen Is "For the Birds"

> When Heaven is about to confer a great office on you, it first
> exercises your mind with suffering and your sinews and bones
> with toil.
>
> Mencius (372–289 B.C.E.)

> The sound of the bell. The chirp of the sparrow. It's through
> these that one meets the true source. Seeking it someplace
> else is a deluded waste of effort.
>
> Ch'an Master Fenyang Shanzhao (974–1024)[1]

More people identify Peter Coyote as an actor and narrator
than as a writer and long-time Zen practitioner. His recent
article in the *Shambhala Sun*[2] dramatized the very hard time
he had coping mentally and physically during rigorous Zen
retreats. [ZB: 138–140] On the other hand, retreatants under
stress sometimes do experience "fringe" benefits. Hence, the
old maxim: "No pain, no gain."

One example Coyote cites was the week-long *Rohatsu* retreat when he was 68 years old. During this annual, worldwide celebration of the Buddha's enlightenment, he was
focusing his attention on a koan. He had condensed it into
the simple question: "What is it?" On the sixth day, late in
the afternoon, he had just stepped outdoors and was taking
several paces to begin the next period of rapid walking meditation. Suddenly, a bird nearby started to shriek: "Eek! Eek!
Eek! Eek! Eek!" Startled, he heard these cries as "It! It! It! It!
It!" in a way that seemed the direct answer to his question.
One step later, and he dissolved into *kensho-satori*. Every old
boundary vanished that had separated Self from other.
Nothing remained of his former fearful, defensive Self.
Nothing remained that *had* to be done. An all-inclusive

awareness perceived this domain. It had no physical location, was "inseparable from the entire universe," and was "perfect, without time, eternal."[3]

These selfless insights were authoritative, yet he was fortunate soon to be graced by the next understanding: nothing about this event was "all that important." The understanding that awakening is "nothing special" is a useful Zen attitude to develop. It throws cold water on any sense of attainment. [ZB: 636–637; SI: 93–94]

Another "Avian Zen" Story

Multiple examples in world literature and in the lore of Zen Buddhism document the fact that birds can trigger an awakening of consciousness. [ZB: 452–460; ZBH: 52–64] Recent further confirmation occurs in Bernard Leach's biography about his close potter friend, Shoji Hamada (1894–1978).[4]

Hamada's fame as a world-class potter was such that he was also celebrated in Japan as a rare "Living National Treasure." He learned from his father, at age 14, how to sit in meditation with his legs crossed, Zen style. One of his early mentors, Sozan Iseki, was a serious Zen student. Hamada's ongoing sense of composure was without parallel in Leach's experience, a friend who had *never* seen Hamada angry.

Starting in 1920, Hamada helped Leach to establish the world-renowned pottery at St. Ives in Cornwall, England. On weekends away from clay on wheel, Hamada often went off on solitary hikes along the clifftops. One day when he was 27, he lay down resting for a long time on a slab of stone on the cliff. Unexpectedly, a wild cuckoo cried out near his head. Startled by the loud call, he suddenly "*knew*." "Ah." …

Could mere words really describe this awakened state of consciousness? No. They would be only "a betrayal of that experience." These words point to *ineffability*.[5] [ZB: 542–544]

Yet, how can we appreciate the depth of Hamada's realization? He states that "my real life" began at this pivotal moment. Indeed, he said, this event was the "central core" of his life story.

Hamada did leave two hints that could help interpret some other, inexpressible sequences later during that peak experience. One was: "I thought, 'I am here. I know it!'" The other was his realization that the cuckoo on the cliffs had taught him that "the object never exists without the subject."

The words "I thought" can be interpreted as pointing toward the reentry of soft word thoughts during the late transitional phase of *kensho*. These are the few seconds when a diminutive i-me-mine begins to re-emerge. Now it confronts the stunning fact: that old, ever-present *I-Me-Mine* Self had just totally *vanished*! Only at this moment—and from this novel, remindful-of-a-whole lifetime perspective—is it possible to comprehend the incredible: one's former dominant thinking Self had previously been inserting its own tentacles of subjectivity into *every object* in the environment! (chapter 4, appendix G). [ZBR: 260–265]

Commentary

As we continue to define the sequences through which a daily Living Zen practice cultivates the soil of enlightenment, it is now time to recall five words that Nanrei Kobori-Roshi once said to me four decades ago: "Zen is closest to poetry." Yes, Zen meditation can become painfully hard on one's joints, muscles, and psyche, as Peter Coyote has described in detail. Yes, serious Zen aspirants can come near their personal tipping point, almost ready to abandon meditation for these and other excuses. Zen practice during retreats then becomes a rigorous test of one's capacity to endure and ultimately to prevail. [ZBR: 126–134]

To prevail into *what*? Many lines of evidence emphasize that major powers are inherent *if* the next state's precipitous arrival is characterized by an acute, inner *silencing* of the Self. [ZB: 367–370, 633–636] The converging evidence suggests that our usual, heavy left hemispheric over-investments in personal language obstruct our dropping into states of *kensho-satori*. [SI: 150–152; ZBH: 30, 148–150, 155–156] Then how can a Living Zen practice of meditation till the soil so that the seeds of these advanced states can emerge? By both silencing the discursive clamorings of this old pejorative Self *and* opening up their allocentric alternatives. [MS: 169–177]

This conclusion opens up some new possibilities for the next topics. Perhaps some random encounters with deep poetic metaphors and with the attention-capturing stimuli that wild birds provide could suddenly play positive, deepening roles for meditators on their Path of Zen. The next chapters illustrate this point. They describe the progress of a young man immersed in poetry who chose to endure the rigors of primitive travel outdoors on foot in seventeenth century Japan.

Living Zen *outdoors* is an integral part of the ancient Zen training process, not some "extracurricular" activity. [MS: 52–60] It was no accident that this young man's personal encounters with Zen evolved into poetry and journals that would celebrate the natural world. Such a personal return to one's deep elemental roots in Nature expresses a normal, instinctual development on the ordinary Path of Living Zen remindfully. [ZB: 664–667] Such an instinctual trend is too important to be viewed as limited to literary issues alone. The ways human beings actually shift toward *eco*centric concerns, away from the dominant *ego*centricity that now prevails, become crucial to ensure that the biosphere survives during our tenure on Planet Earth.

22

Basho, the Haiku *Poet*

What makes Basho one of the greatest of the poets of the world is the fact that he lived in the poetry he wrote and wrote the poetry he lived.

R. H. Blyth (1898–1964)[1]

Poetry? What does it have to do with Zen? Well, let's begin with the old Japanese expression: "Poetry and Zen are one" (*Shizen ichimi*). Maybe you might not want to write poetry yourself. Yet, by the poet Marianne Moore's criteria, you could still be a person who is *interested* in poetry.[2] She wrote that you could be this interested person *if* you demand two things: (1) the raw material of poetry "in all its rawness" and (2) genuine poetic expression.

If you are also a minimalist, these expectations might lead you into the spare verses of *haiku*. And then, inevitably toward Basho, whose frog-"*Plop!*" is still being heard around the world.

Can anything *new* be written about Basho? Even two decades ago, Ueda found that the "huge accumulated mass" of daunting Basho scholarship was already intimidating. He concluded that the *non*-Japanese readers of Basho would be the ones who now had the greatest opportunity to make novel contributions.[3] Did reading words such as these encourage the present attempt to stimulate further discussion?

No. In the interest of full disclosure, this essay's motivations began over eight decades earlier. I grew up in the city. Not until I visited my uncle's farm at the age of nine did I first hear this distinctive "*Plop!*" It arose from a region several yards away. Here, a pool lay hidden beyond the bend of a gentle meadow stream. *What* caused that curious sound? I went there, saw nothing, remained mystified. Days later, I finally *saw* a frog leap into this water. Simultaneously, I *heard* what happened. This sudden comprehension resolved the mystery. The deep impression it left [SI: 154–158] remains memorable to this day. [MS: 156–163]

In these next pages we take a neural perspective to examine the evidence, pro and con, that Zen sensibilities came to mature in the later life and poetry of the man born as Matsuo Kinsaku in 1644.

Poets as Literary Revolutionaries

Large gaps still remain in our understanding of this man later known by his pen name, Matsuo Basho. We do know that he accomplished much during a short life of only five decades. For example, earlier centuries of Japanese poetry had given voice to only the ordinary croaks.[4] In the spring of 1686, Basho became the first poet who immortalized the raw, distinctive, *frog-water* sound. Moreover, before he arrived, the 5–7–5 form of Japanese poetry "had largely been an entertaining game."[5] Soon, Basho would help lead the literary revolution that rescued the *haiku* from the artificialities infiltrating the sonnet in Western verse.[6]

Sounds travel far. Echoes relay the message. A century and a half later, a longer poem was also becoming well known, on the other side of the world. Its author was Ralph Waldo Emerson (1803–1882). Emerson, like Basho, was not only a keen observer of Nature but also a writer who helped to inspire a literary revolution in his own country. Emerson's poem ("The Concord Hymn") dramatized no natural sound. Instead, it echoed the first shot fired "by the rude bridge" at Concord, Massachusetts. That pivotal sound, back in the spring of 1775, was also far-reaching in import. Indeed, it was the herald of revolutionary trends still going on in distant lands today, of echoes that still keep being "heard 'round the world."

Reginald Horace Blyth Examines *Haiku* and Their Relation to the Practice of Zen

In 1963, Blyth published the first of his two major volumes on *haiku*. It surveyed its centuries-long history and devoted to Basho a section of 24 pages. Following this were 75 pages about the poet's closest disciples and others who were members of the already large Basho school. Given the old saying, "Poetry and Zen are one," and because Basho was a lay student of Zen Buddhism, Blyth sought answers to two questions: Did Zen influence Basho's poetry? And, does poetry help us understand Zen? These chapters continue this quest.

Blyth began this analysis with two cogent quotations by Bernard Phillips:[7] "Zen practice is at bottom the act of giving one's self, of entering wholly into one's actions." ... and "One must enter into the practice and become one with it, so that it is no longer an action performed by a doer who is external to the action." Then, Blyth used nine words to distill his own version of the intimate relationships among Zen, poetry, and *haiku*: in Zen, he said, "Doer is deed." In poetry, "Word is thing." In *haiku*, "Meaning is sensation."[8] Pause ... Allow these nine words to incubate ...

It now becomes essential to ask why Blyth also ventured one crucial statement. Why did he conclude that when Basho's *haiku* flowered during his later years, this later period marked "the beginning and the end of the poetry of the thing in itself"?[9] What *is* "the thing in itself"? How could Basho's poetry be oriented toward a concept this abstract? Please bear with the explanation in the next paragraphs. They continue a major Self—other theme that began in chapters 4 and 5.

The phrase *Ding an sich* is associated with Immanuel Kant (1724–1804). It is beyond our present scope to explore every philosophical implication of these three German words. However, the phrase does help to conceptualize how the noumenon (or suchness) might coexist in its relationships with one's personal Self. [ZBR: 361–371] Concepts that distinguish between "the other" (outside) and the Self (inside) not only are grounded in neurobiology, they become fundamental in understanding Zen teachings. So, is there some simpler way to understand Blyth's nine words and his conclusion? Is there also a way to grasp how *we* and Basho— stuck inside our *own personal* human frame of reference—are the "doers" responsible for *re*creating each meaningful "sensation" that arises when stimulus energies enter us from each "thing" out there in that other world outside our own skin?

These intricate issues become more tractable when they are approached from a simplified model of brain anatomy and physiology.[10] From this practical neural perspective, the proposal is that our normal consciousness is based on its two major *frames of reference*: Self and other. Our perspective from this first compartment can arise only with reference to the internal axis of our omni-Self. At the same time, that other compartment represents—*anonymously*—the exterior world as occupying the space "outside" of our skin. These two Self—other categories are implicit in the blend that begins to interpret our everyday notions of "reality" (chapter 4) It's useful to review these concepts and their implications because they are hard to grasp and harder to remember.

First notice how much extra weight we give to our own subjective *Self*-centered constructs. The mergers in our sense of "reality" remain *asymmetrical*, unbalanced. Why does our basic *ego*centric bias remain so dominant, so intrusive? Because this first, Self-centered frame of reference emphasizes its *two* huge subsets of distributed functions. The Greeks called them our *soma* and *psyche*. To each subset of our Self, we assign the highest of priorities.

Our *soma* is the source of the normal egocentric bias that is easiest to recognize. It's *tangible*. We refer all its percepts back to the physical axis of our own body (our *soma*). (You can demonstrate this by touching and pinching your own arm. Notice how you feel your sensations converging on *you*.) In contrast, the second subset in our bias is not so easy to pin down. These Self-centered, subjective thoughts and allied mental operations of our *psyche* are *intangible*. (You can't touch any thought, although the longings and loathings within your psychic Self express emotions that *can* leave you feeling uncomfortable.)

The second frame of reference offers a sharp contrast. Its function is to represent any "thing" in that huge world *outside* our skin. This *other* version of "reality" arises within subordinate networks. They receive a lower priority. Still, they may seem more objective. Why? Because their covert functions emerge *anonymously*. What happens in the brain when this second category of anonymous neural codes perceives signals coming from things *out there*—from items out in that *other* world of space outside our skin? It assigns those "things in themselves" to "*their*" own, 3-D spatial relationships, *not* to lines of sight that converge through the lens on our own Self. Such an "*other*-centered" kind of attentive processing is called *allo*centric. In Greek, *allo* means other.

Ordinary consciousness blends both ego- and allo-centric physiologies seamlessly. Diagrammatically, this strong Self-centered bias of our psyche and soma can be represented as: ego >>> allo.

This neural perspective helps clarify why Blyth found that Basho's later poetry was so oriented toward "the thing in itself." In short, Blyth recognized that these later *haiku* tended to drop Basho's old *subjective* frame of reference. No longer did it point back toward some human author. Instead, these *haiku* were opening *outward*. They were perceiving *other* things, just as *they* existed, more objectively, outside in that world at large. *Out there*, these other things drew attention to *themselves*. Those former notions of *My* Self were finally taking a subordinate role. Now, allo >>> ego (J:

muga). In this unattached form of poetic expression, authentic *haiku* ventures toward the ineffable, the inexpressible. [ZBR: 358–361]

Comment

Suppose you and I also *let go* of our unfruitful subjectivities. Suppose we not only drop the maladaptive aversive responses of our fearful Self but also let go of every excessive longing that had dominated our habitual approach behavior. Could this liberate some more positive, innate options, some mature behaviors formerly buried among the instinctual networkings of our brain? D. T. Suzuki called these our basic, "native virtues." [ZB: 648–653] Multiple affirmative potentials are inherent when we do let go of our precious Self in this liberating manner. These potentials include authentic forms of compassion.

Meanwhile, we notice the restraint with which Blyth deferred his own English translation of Basho's overly familiar *haiku*. Not until the final chapter of his second volume would readers finally see gathered all the lines of the *haiku* they've been waiting for, their luster undimmed by earlier, partial translations.[11]

> An old pond;
> A frog jumps in:
> The sound of the water.

Robert Aitken Examines the Question: Was Basho's Maturation as a Person Expressed in His Poetry?

When the United States entered World War II, Robert Aitken (1917–2010) was a civilian working on Guam. The Japanese interned him at a camp for civilians in Kobe. There, fortunately, he happened to find that one special co-internee who would become "like a father in helping me to find my life path."[12] Inspired by Blyth, Aitken returned to Japan after the war, embarked on the Path of authentic Zen Buddhism with Yamada Koun-Roshi and others, and matured into the founding leader of an extended Zen sangha based in Hawaii, a socially engaged Buddhist roshi, and an internationally known author.

In 1978, in *Zen Wave*, Aitken developed his own analysis of Basho. It was informed by a rare combination of key attributes: a thorough first-hand Zen perspective as both student *and teacher*, a working ability to translate Japanese, and a scholarly understanding of the relevant literature. How did these capacities enable Aitken to envision the sound-image of a frog entering water? As an event resonating within the immensity of an *Indra's*-net-like universe. Viewed in the context of this coreflective universe, we humans are simply coparticipants along with what we hear and see, and all other phenomena. Aitken could regard the scope of this boundless universe as "a complementarity of empty infinity, intimate interrelationships, and total uniqueness of each and every being."[13]

Aitken devotes his first chapter to The Old Pond *haiku*. He comments that its fame, after more than 300 years of repetition, had understandably become a little stale among Japanese-speaking people. However, he also points out one advantage of this situation: those who read it in English now have "something of an edge in any effort to see it freshly."[14] Can today's search for a fresh, neural perspective take us toward such an edge? It can when we overcome our initial resistance to unfamiliar words in Greek and German and new concepts in neurobiology.

Speculating, Aitken envisions the setting for this *haiku* — a large overgrown pond in some public garden. He imagines Basho there, "lost in the samadhi of an old pond," during twilight in late spring.[15] A caveat: "*samadhi*" is open to several interpretations. [ZB: 473–478] Were we to constrain one of its current meanings to a state of one-pointed absorption, then *samadhi* would usually refer to only a preliminary state of consciousness. This absorption would have no lasting potential to deeply transform a person. Here, however, Aitken prefers a broader interpretation of the word, one that could have the potential to transform.[16] When he writes that Basho "changed with that *plop*," he invites readers to conclude that *this* change represents a deeper, alternate state of consciousness. Such a change—in the direction of *kensho-satori*—would signify that Basho had dropped into a much more advanced state of awakening (chapter 2).

Why does Zen Buddhism emphasize these extraordinary states of insight? Because they can help us change more than our usual "mind." In fact, they can help transform our overconditioned, unskillful *traits* of character (chapter 3). Other assets of Zen training are being increasingly recognized (chapter 1). When covert resources of attentive processing are retrained and flexibly applied, they can contribute to our potentials for creative behavior in ways we're not aware of (chapter 7).

But what about a mere sensate stimulus (one perceived as giving rise to a "sensation" in Blyth's triad of words)? Can it spark deep, hidden links to "meaning," enabling even a single "*Plop!*" to precipitate a major alternate state of consciousness? That mysterious sound of a frog certainly kindled *my* childish curiosity. But could it also strike a chord that would trigger a grown man into such a major physiological reaction?

Can a Sensory Stimulus Precipitate *Kensho-Satori*? Remote Historical Evidence

Many examples in the old Ch'an and Zen literature testify to this fact: triggering stimuli—*especially* sounds—can precipitate major alternate states of consciousness. Plausible neural explanations suggest that when such stimuli strike a sensitized brain, they instantly become salient in ways that cause deep normal shifting mechanisms to topple over. [ZB: 241–242, 452–457; ZBR: 303–306; SI: 109–117, 189–193; MS: 76–77]

Aitken-Roshi points to the example of Xiangyan Zhixian (J. Kyogen; died 898).[17] As this monk was sweeping around the grave of the National Teacher Nanyang Huizong (675–775), his broom suddenly dislodged a tile stone. It flew on to strike a nearby bamboo stalk. The resulting *"Tock!"* precipitated Xiangyan into a deeply enlightened state of consciousness. This state resolved the essence of a certain koan about the deep meaning of one's "original face," one that he could not penetrate previously. The next chapter offers further comments on this koan.

Did a Triggering Sound Stimulus Actually Precipitate a State of Awakening in Basho? More Recent Sources of Information

In 1759, Zen Master Hakuin Ekaku (1685–1768) published one of his paintings that portrayed Basho with his "old pond" poem. Hakuin's accompanying idiograms (in Chinese) suggested that this master poet had "dropped off mind and body" when he heard a frog jump into the water.[18] This phrase echoes "dropping off body and mind," a phrase associated with awakening when used earlier by Master Dogen Kigen (1200–1253). [ZB: 576] What does such a letting go of one's somatic *and* psychic Self imply? It means entering the advanced state of awakening, during which *both* of these egocentric subsets vanish from the field of consciousness. Notice again that this dropping off of one's Self-centered body and mind still spares the *other, allocentric* frame of reference. Now *all things in themselves* can dominate the entire field of consciousness.

Master Hakuin's painting was published only seven decades or so after Basho wrote his "old pond" *haiku*. Hakuin was an exacting judge of character. Was his brush just playing with the early stirrings of the legends that some might later refer to as part of the "Basho myth"?[19] Or, did Hakuin already have access to more facts than we're aware of—to details and dates that convinced him Basho really *did* undergo such a dramatic change when he heard that frog-water sound?

In the last century, Daisetz T. Suzuki added a different version of the frog "story."[20] He did not identify its source. Nor did he specify precisely where, when, or why the details of its several constituent events could have taken place. However, the "story" narrates a conversation that Basho was said to have had with his teacher. On this occasion, he was visited by (or visited) his Rinzai Zen teacher, Butcho (1643–1715). Basho had studied Zen under this teacher's guidance at Chokei Temple in Edo (sometime) between the years 1673 and 1684[21] or 1681 and 1684.[22] As this story begins, Butcho asked his student-friend the kind of conventional opening question that has Zen overtones: "How are you getting on these days?" To his teacher, Basho replied: "After the recent rain the moss has grown greener than ever."

Suzuki then says that Butcho "shot a second verbal arrow" to probe the depths of Basho's spiritual understanding: "What Buddhism is there even before the moss has grown greener?" This kind of question addresses the "Big Picture." It is equivalent to asking a student: Where are you on the Path in particular relation to the universal, timeless, undifferentiated absolute?[23]

At this point, in Suzuki's account, Basho is alleged to have answered: "A frog jumps into the water, and hear the sound!" (*sic*) Suzuki suggests that [at some moment] when "Basho himself was altogether effaced from his consciousness," this water sound was heard "as filling the entire universe."[24] When a witness tries to describe such an advanced state, a state that *both* effaces the Self *and* opens into such inclusive allo phenomena, it is important not to dismiss these wordy attempts. Nor should either of these two overlapping categories of direct experience be confused with metaphysical notions, metaphors, symbols, or with simple poetic license of the "'round the world" kind. Instead, when a sound seems to resonate this far beyond its usual boundaries, it can have authentic neurophysiological correlates.[25]

Suzuki goes on to inquire: What made this "old pond" *haiku* of Basho so "revolutionary?" Why did such a revolution mark "the beginning of modern *haiku* poetry?" We must read *between* the lines, he says, because *here* is where Basho's revolution begins. Where? In the deep structure of meanings hidden in the background of this verse. It originates in Basho's "insight into the nature of life itself, or into the life of Nature. He really penetrated into the depths of the whole creation, and what he saw there came out as depicted in his *haiku* on the old pond."[26]

To help clarify what such a deep insight means, Suzuki then explains that a *haiku* does *not* express ideas. Instead, a *haiku* "puts forward images reflecting intuitions. These images are not figurative representations made use of by the poetic mind." Instead, "they directly point to original intuitions," and "are intuitions themselves."[27]

Suzuki postulates that Basho had a direct experience that arose from such intuitive depths.[28] Such a first-person experience "was given an expressive utterance in his *haiku*."[29]

Suzuki presents his own translation. Its two exclamation marks signify his points of emphasis[30]:

> The old pond, ah!
> A frog jumps in:
> The water's sound!

He explains that Basho added *later* the phrase that would describe the setting in his first line. By adding this phrase, "The old pond," Basho was able to complete the requisite 17 Japanese syllables for this *haiku*.

In 1991, Makoto Ueda imagined that the last two lines of Basho's poem were the result of an "indescribable sentiment" that had "floated into his mind." He placed Basho at this ineffable moment at his *riverside* hut in the north of Edo in that spring of 1686.[31] He indicated that a student was then by Basho's side. This student, Takarai Kikaku (1661–1707), had suggested that "the mountain roses" might be a good first line. Instead, Basho preferred the much more appropriate phrase, "The old pond."[32]

By now, readers will recognize how many details are ambiguous. Did this frog-water sound enter from some remote experience in Basho's distant past, perhaps at some *other* pond? Or did it represent some relatively more recent experience, perhaps even at the nearby Sumida River? We might also be wondering why Basho would have uttered the strange words "rain" and "moss," let alone "*hear* the sound," in this (undated) dialogue with his teacher. Was this in reference to some earlier incident(s) they both knew about? Or was it related to a particular recent, overnight rain that they had both just shared?

In this century, the well-known *haiku* poet Jane Reichhold collected 1012 of Basho's successive *haiku* in one publication. The "old pond" *haiku* enters only as number 152 in this extensive series.[33] Therefore, the vast majority of Basho's *haiku* still remained to be written.

In Japanese:

> *furu ike ya*
> *kawazu tobikomu*
> *mizu no oto*

In Reichhold's spare translation:

old pond
a frog jumps in
the sound of water

Reichhold identified 33 different techniques that Basho used in his many *haiku*. She presents examples of each. In the eighteenth technique, we realize why Basho's last two lines contain a word image that could require us to make a perceptual shift. Now we understand what these words could mean in the original Japanese.[34] They could suggest that the *frog jumps into the sound of water.*

So, does this frog's *"plunge!"* into water capture the witness's vision in those same milliseconds that its *"plop!"* also captures hearing? If so, then perhaps these *two* sets of incoming sensory signals are *mutually reinforcing triggers* for some neural events starting as low as in the colliculi of the midbrain. [ZB: 241–242] The resulting phenomena might coalesce in the lived experience of a poet-witness, a person who later could only allude to those deep shifts into other percepts that had also suddenly become egoless. These other shifts have the capacity to efface the Self *and* to dissolve both the psychic and somatic boundaries that would separate this Self from the outside world. At this instant the witness's spared allo- resonating perceptions could expand into a vast field of undifferentiated oneness. [MS: 156–165, 169–177][35, 36]

We hope to welcome any new factual evidence forthcoming from Japan. If new facts could confirm that such a selfless, timeless, fearless coherent state had actually opened up in Basho's consciousness (sometime, somewhere), then they would support the proposals advanced earlier by authorities such as Hakuin, Suzuki, and Aitken. Yet, Basho's cryptic words have aroused our curiosity. When, where, and why could any such "rain" and "moss" have entered into his response to Butcho's question?

Basho's Ongoing Quests: A Traveler on Both an Interior and Exterior Journey

In this detective work, our own quest could certainly be called a "cold case." However, Basho's remarkable travel journals disclose countless items of information that help us understand him as a poet and as a person. So, it will help us to travel along with Basho on his literary journeys. Let us choose, only for comparison, a later incident in 1689 (table 24.1). Basho described it in his account of the long journey through the northern provinces.[37, 38]

North of Nasu, he found the moss "dripping" at the higher wet elevations. Why was Basho searching up here for one special site? Because this was the place that, in the *past*, had served as the hermitage for his teacher. Butcho had chosen to live at this site when he had gone off on an earlier solitary retreat in the mountains. Moreover, after Butcho had returned from this retreat, he had then described his experiences there to Basho (at an unspecified number of times and places during the subsequent years).

Located up the hill from Unganji Temple, this special place was only "a tiny hut atop a boulder and built into a cave."[39] There, on the surface of a nearby rock, Butcho had once used a stick of pine charcoal to inscribe this poem. Its words would long since have been weathered away:

> A grass-thatched hut
> less than five feet square:
> regrettable indeed
> to build even this—
> if only there were no rains

This was a crude, makeshift, hut-cave. Indeed, it was a site so primitive that when Basho left his own parting verse pinned to its pillar in 1689, he included a significant avian point of reference: "even woodpeckers don't damage this hut." We discover how easily each of these two Zen practitioners could infuse resonances of meaning into their verses about daily life events.

Basho's whole entry about this hermitage in his journal throws wide open some new windows for fresh interpretation. Its cluster of intriguing items points toward key dates in the earlier close relationships between the two men. Knowing these shared dates could help us to clarify otherwise ambiguous details in the stories and *haiku* discussed in this and the next chapters. For example, why did the conversations between these two Zen friends, teacher and student, include wet, rainy, mossy conditions? Now they also involve impermanent lines of verse left by transient human beings, verses that would inevitably be worn away by rain, wind, and regrowing moss.

Among the new options now are these possibilities: (1) Basho, the Zen student, might have been *presenting* his teacher with metaphors from private conversations that they had shared about *this* hermitage in the past, (2) Basho might have been using words like *"hear the sound of the frog entering water,"* alluding to this prior incident, in response to Butcho's second question. In the time-tested intimacy of such a dyadic relationship, a few select cues in a Zen student's response, when reinforced by corresponding body language, suffice to deliver the requisite analogy in cryptic form, (3) Basho might have been alluding (in Suzuki's story about the exchange with Butcho) to a different special site, an earlier one that we know they shared *at the same time*. In the next chapter, we'll take up this important episode when the two men met at Kashima during the rainy fall of 1687. Barnhill indicates that a growth of "dew-wet moss" had also covered the sacred foundation stone there at its nearby Shinto Shrine.[40]

To Summarize

Yes, something major happened after Basho heard the distinctive frog-water sound. When and where did this incident first occur? We need more facts. An iconic *haiku* was the indirect effect of this sound. The sound and its *haiku* keep echoing throughout the whole wide world of literature. However, the historical evidence *currently* available in English and assembled above still seems circumstantial, perhaps hearsay, and open to alternative interpretations. For example, we cannot conclude, beyond doubt, that just one special "*Plop!*" must have been *the only* trigger that could directly have caused Basho to drop into his one and only enlightened state of consciousness, an isolated state that in itself would have been the sole cause for him to mature during his 40s.

Which other events, *after* the spring of 1686, could have significantly influenced Basho and his *haiku*? In the next chapter, our quest leads us again to select certain narrative incidents from Basho's journals. They will be supplemented by a few selections from the next 860 *haiku* that he would write in those eight short, productive years before he died in 1694. These examples provide overt contrasts. They allow each reader to decide whether any given event, at a particular time and place, appears to be of major, minor, or negligible import.

Then, in chapter 24, the plan is to document a recurrent avian theme. It is a topic that would increasingly attract Basho's attention and be actualized in his later *haiku*.

Basho's States of Consciousness

The senses, particularly sight and hearing, provide the most basic link between the outside world and the activities of the mind.

Shodo Harada-Roshi[1]

The poet's voice need not merely be the record of man, it can be one of the props, the pillars to help him endure and prevail.

William Faulkner (1897–1962)[2]

Did Basho efface the Self during one or more later states of consciousness? If he did, how was this expressed in his later poetry and in his other behavior?

Aitken's analysis suggested that Basho's *successive haiku* manifested a developing maturity during the eight years before he died. This maturity mirrored his own development as a person.[3] Indeed, based only on his readings of these *haiku*, Basho's verses appeared to "point precisely" to "metaphors of nature and culture as personal experience."[4] Hamill's and Reichhold's reviews agree that Basho showed increasing evidence of *wabi, sabi,* and *aware* in keeping with this deepening of personal and literary maturity during his later forties.[5, 6]

Could one early episode in this evolution toward maturity have occurred in the *fall* of 1686, not in the spring? Ueda dates this potential incident to October 2, 1686.[7] On this evening, Basho and a few of his students were enjoying a moon-viewing party at his hut north of Edo. There, at a pond, Basho became impressed by viewing the full moon and then composed the following *haiku*[8]:

Meigetsu ya
ike wo megurite
yo mo sugara

This translates as:

The autumn moon!
I walked around the pond
all night long.

Yes, it would be unusual behavior if a person actually did stay awake during an entire night while circling a pond repeatedly. Aitken suggests the possibility that when "Basho wandered about the pond in the moonlight" an awesome awakening might have opened into a state of "suchness and emptiness of all beings and all things." We don't know how "long" this actual walking lasted. Its duration could have been exaggerated. Again it is uncertain which pond was involved in this incident.

In Reichhold's translation[9]:

full moon
walking around the pond
all night

She observes that its wording permits the *haiku* also to be read as consistent with a very different idea—that the *moon* "walks" around the pond during the night. She uses this same ambiguity as an example of Basho's twenty-sixth *haiku* technique. By "hiding the [human] author," the poet is allowing Nature to express itself. This *haiku* can serve as another illustration of Basho's personal trend in the direction toward *allo*-perception and away from *ego*-perception.

Oseko suggests that the episode is interpretable as the poet's being unaware of the lapse of time.[10]

Earlier in 1687, Basho begins to write of skylarks singing, but first he writes a *haiku* about lightning. He sends it off to his bright, talented disciple, Rika, who lives in Edo. Rika is special. He is the disciple who gave Basho the tree that would become his pen name.

Representative translations of this 1687 *haiku*, by Barnhill[11] and Reichhold,[12] respectively, take the following forms:

Lightning	a flash of lightning
clenched in his hand:	your hand takes in darkness
torchlight in the dark	a paper candle

Could such a terse verse be an oblique hint to Rika, reminding him that a particular Zen story in the *Blue Cliff Record* conveys personal implications? This familiar story narrates the major nighttime transformation of a bright, scholarly, contentious monk who lived back in the Tang Dynasty. In Chinese, his name is Deshan Xuanjian (780 or 782–865). In Japanese, his name is Tokusan.

Young Deshan became memorable for two major reasons.[13, 14] First, because—to his surprise—Ch'an Master Longtan Chongxin suddenly blew *out* the lighted paper torch that had previously enabled them to see in the darkness. This was a turning point, a big, unexpected sensory *change* (chapter 12). It was a visual trigger so striking that it abruptly precipitated Deshan into *satori*. Second, because the next day Deshan burned all his earlier opinionated scholarly commentaries on the *Diamond Sutra*. Why? Because now—having just been transformed in the depths of *satori* the night before—he had realized that all his earlier writings were inconsequential.

A lightning strike is one ancient metaphor for the flash of *prajna's* insight-wisdom. The flash of lightning carries an implication that life itself is also transitory. Life would also seem to have been transitory for Basho, given the few years he was left to live.

On balance, it is the *next* happenings, starting later in the autumn of 1687, that appear increasingly relevant to Basho's affinities for Zen and to the trajectory of his personal and literary development.

Later Events at Inkyoji Temple in Autumn 1687; The First *Haiku* (Number 316)

On August 14 Basho began a journey to view another harvest moon. This trip took him into the lake country around Kashima, 110 km to the northeast of Edo. His two traveling companions were Kawai Sora (1649–1710), a young student and neighbor,[15] and Soha (d. 1512–16), a Zen monk of Teirinji Temple who lived near Basho's hut.[16] Unfortunately, it began to rain on the afternoon of September 21, the date they had anticipated seeing the full moon. Basho spent that rainy night with his old friend and Zen teacher, Butcho.[17] Although this rainy night graced Basho with only fleeting glimpses of the fugitive moon, it did stimulate him to write two significant *haiku*. The first of these in Reichhold's[18] series is:

tera ni nete
makoto-gao naru
tsukimi kana

In translation:

sleeping at a temple
with my true face
moon viewing

The key operational words are *makoto-gao naru*. *Makoto* means really, truly, authentically, sincerely. *Gao* means face. *Naru* has several intriguing meanings. One of them is to resound, the way a bell resonates when it is struck. Other meanings include to become, to grow, and to turn. Therefore, individual readers from different perspectives are being invited to interpret this *haiku* at various levels.

Addiss[19] translates this *haiku* as:

> staying at a temple
> with my own true face
> I gazed at the moon

He interprets the second line of this moon-based *haiku* as recalling the deep spiritual awakening that realizes one's "original face." The basic issue is embedded in koan number 23 of the *Gateless Gate* (*Wu-men-kuan*). In this koan, the legendary sixth Zen Patriarch, Dajian Huineng (638–713), poses the key question to another monk: "What is your original face, before your parents were born?" [ZB: 540–542]

If this *haiku* were to point toward the immediacy of an awakening, then its translations might be further "lightened." A first step could invoke the ancient aesthetic principle: *less is more*. How could this moment be recast into a selfless mode of non-doing? By dropping all Self-referential verbiage (my, I) from the last two lines. The second line would then barely hint that this rare state, graced by the impression of authenticity, is arising from the instinctual networks of an egoless brain. No such moment is gazed at, or "owned," by any Self inside. It is a moment *oned* (*ichinyo*).

A further condensation could begin by translating the final *kana* of the third line to suggest a heightened sense of closure. Placing an exclamation mark at the end could also serve to convey the stark internal vacancy that remains when such a long-awaited full moon finally does appear.

So:

> At the temple,
> original face;
> cool moonlight!

Zen literature contains multiple examples of similar extraordinary realizations. They strike in a flash, quick as lightning. Lacking every trace of the old personal Self, they reveal an arctic, empty quality to the scene as it is embraced by moonlight, in an absolute vacancy of time (achronia). [ZBR: 434–461]

Reichhold concurs in the interpretation that such an image of a "true face" in Basho's *haiku* refers to "one's original being, before emotions are added to one's life."[20] This cool Zen perspective is absent if a translation were to limit *makoto* to mean only "a serious look."[21]

In brief, today's neural implications of "original face" mean that one's limbic networks stop infusing their excessively overconditioned emotional burdens. Those old "hot fires" of greed, hatred, and delusion are briefly extinguished after having generated much previous suffering in oneself and other persons. [SI: 90–94] This cool quality in awakening had long before entered other Asian verse, yet no such hint had entered this *haiku* in its original form.

When Basho was at Inkyoji temple during portions of these last two days, he was interacting with Butcho, his former Zen teacher.[22] This time would have presented opportunities for the two men to be reminded of those pivotal experiences they had shared during their many years together. Such a renewal of old memories might have enabled their hours in the atmosphere of this Buddhist temple to evolve into a sensitizing setting, one that had the potential for some kind of "opening" to occur.

Yet strong cultural traditions could prohibit some poets from mentioning their own private experience with such a well-known koan from the *Wu-men-kuan*. This koan (case 23) is noteworthy for its multiple allusions. They hint at an alert state of selfless insight, a state emptied not only of every word-thought but also of all polarized emotions and all dimensions of time. Indeed, many Zen students who have wrestled with the surface words of this opaque koan would never think of "breaking ranks" by referring to some of its layered implications unless compelling reasons existed for doing so. [ZB: 536–544]

Question: Was it an impropriety for Basho to use words that might leave the impression he had actually experienced an advanced state comparable with a "true (original) face?" Not necessarily. Variations of the phrase *original face* are not uncommon in Zen dialogue (*Honrai-no-memmoku*). To mention these words need not imply that the trainee was now working with this particular koan at the present time or had been in the past.

Another fact of historical interest could also tilt Basho toward taking a waiver on such cultural taboos. He greatly admired the poetry and the traveling life style of Saigyo Hoshi (1118–1190), a Shingon Buddhist monk. In Saigyo's era it was customary for Japanese monks and the Buddhist laity to engage in the formal meditation practice of moon gazing (*tsukimi*). In fact, five centuries earlier, Saigyo had also broken ranks. He had inserted the word *satori* into one of his own longer poems.[23] Its verses can be translated as:

> In the deep mountains
> dwelling in the moon
> of the heart-mind
> this mirror sees
> *satori* in every direction.

Ample evidence exists that Basho was also attuned to the innermost feelings that he would discover during deep mindful introspection. For example, during the following summer of 1688, he saw another full moon at the seacoast. That moon did not resonate with those additional deep-feeling tones he had experienced while he glimpsed the moon at the temple back that previous fall on the Kashima trip. His internal sensitivities discerned the big emotional difference in *mood*light that had separated the remarkable autumn moon[!] of 1687 from this ordinary moon at the seacoast in 1688.[24, 25]

The foregoing lines of direct and circumstantial evidence converge on several potential options. At *this* Buddhist temple, on *that* rainy fall night of September 21–22, 1687, Basho *might* have dropped into a major alternate state of consciousness. Yet, in the previous chapter, Master Hakuin, D. T. Suzuki, and Robert Aitken had suggested that a more advanced Self-effacing state had been triggered by the frog-water sound a year or so *earlier*, back at least in the spring of *1686*. Therefore, what other facts add weight to this later rain—moon incident at the temple in the *fall* of 1687? And *if* Basho could have undergone *two* such "opening" experiences in separate years, could that rainy fall night in the Buddhist temple have been the more decisive one, a final straw, as it were? Or was it of only relatively lesser significance? Clues exist.

Supporting Details: Wet Leaves and Raindrops at Inkyogi Temple in Autumn 1687; The Second *Haiku* (Number 317)

That same rainy night in the temple led Basho to compose a second *haiku*[26]:

> a quick glimpse of the moon—
> while leaves in the tree tops
> hold the rain

This second *haiku* includes two important details:

- Now Basho is gazing *upward*, up toward the *tops* of trees.

- *Out* there, when the rain stops, he sees this striking visual image: moonlight glistening off the foliage of *wet tree leaves*.

Haiku 317 is significant for another reason. Its timing corresponds with another noteworthy entry in Basho's daily journal. This description is found in *The Record of a Journey to Kashima*. In Reichhold's translation [27]:

> In the light of the moon, the sound of the raindrops
> was deeply moving;
> our breasts were full, but no words could express it.

Notably, this entry from Basho's journal in the fall of 1687 adds three more supporting details to the first two on the list above:

- Two kinds of stimuli are converging. These stimuli are both visible (moonlight) *and* audible (the sounds of falling raindrops). This combination of sight and hearing is potentially triggering, a point alluded to by Harada-Roshi in the epigraph.

- Basho is infused by a "deeply moving" affirmative experience.

- This experience is also ineffable (indescribable).

On the other hand, Barnhill's translation introduces an alternative concept.[28] His translation of Basho's entry reads: "For a while, I felt peace and purity sweep over my heart … [In] The moon's light and the sound of rain: I became absorbed in the deeply moving scene, beyond what any words can tell."

Barnhill's translation at this point then adds a second entry from Basho's journal. It indicates that Basho identified with the disappointment of the woman who had returned home distressed because she had not succeeded in writing an authentic poem about a cuckoo. Disappointment is not an impression that is part of the usual afterglow of *kensho*. A more likely origin for a sense of disappointment might arise *if*, after having emerged from a state of absorption, the person recognized that this state was *not* one that had deepened further into *kensho*.

Parenthetically, D. T. Suzuki, in the context of *haiku*, does refer to the *haiku* poet, Chio. Only after many attempts was she finally graced with a genuine *haiku* about a cuckoo.[29] This *haiku*, devoid of egotism and artifice, had issued from her subconscious. Suzuki wisely observes that this is the deep realm "where artistic impulses are securely kept away from our superficial utilitarian life. Zen also lives here, and this is where Zen is of great help to artists of all kinds." Did more than one *haiku* poet write about the cuckoo? Yes. Because Suzuki dates *this* poet Chio to 1703–1775, long *after* Basho died.

Full Moonlight Glistening off Wet Tree Leaves

Some night, *after* the rain stops, go out and look up into what occurs when wet foliage is backlighted by a full moon. Examine first the effects created when that familiar bright moonlight enters only through a patch of clear sky. Then compare it with the way this same full moonlight shatters as it glistens off hundreds of wet leaves interposed up in the adjacent treetops. If you're fortunate, you'll now be seeing something rare—hundreds of tiny bright points of reflected light set against their darker, arboreal background. The novelty of such a sight could have captured Basho's attention.

The Sounds of Raindrops

Anyone can hear the familiar *pattering* or *drumming* of rain. Anyone can hear the "*Plop!*" or "*Ker-plunk!*" of a frog. These mundane sounds might seem hardly worth mentioning. Unless, of course, these natural sound energies strike a sensitized person. Suppose that person's hearing and feelings have become keenly aware, that his or her psychic and somatic senses of Self happen to be on the brink of dropping off. Now, simple sound stimuli could trigger the abrupt shift that breaks open into an alternate state of consciousness.

Amid the drumming of raindrops, and then the awesome embrace of moonlight, can even poets easily access words to describe a chest full of emotion? No. Deeper alternate-state experiences drop into otherworldly domains. They resonate beyond reach of language and feel inexpressible to any witness. States of *kensho* arise from networks much farther beyond wordy attachments than do states of absorption. Moreover, a person's spontaneous act of gazing *up* and *out there*—into the distance—can also set in motion deep shifting mechanisms. [MS: 20–37; 72–91] These shifts enhance *allocentric* (other-referential) attentive processing. And—at the same instant—they also drop off one's prior Self-centered frames of reference.

Raindrops that strike a tile roof are not trivial sounds. Notice how a hard rain directs your auditory attention way *up there*, above, to the sounds impacting on the roof *overhead* (as the airplane did for E.G. in chapter 11). With regard to raindrops' triggering potential to raise your gaze, Shodo Harada-Roshi points to their role as potent stimuli that can deeply transform a person's consciousness.[30] In the context of the fourty-sixth case of *The Blue Cliff Record*, he observes that "When we hold on to nothing, we become the sound of the raindrops [*uteki sei*]. The raindrops become us ... Unless we realize this sound of rain that fills the heavens and the earth, we will never know true joy." Notice what these words imply: (1) no personalized attachments; (2) no self/other boundary; (3) self-other *coidentity*, instead.

So, did Basho experience such a state of awakening that rainy night at the temple? Did Suzuki's (undated) conversation between Basho and Butcho actually take place at *that* temple in late September 1687? Clearly, we have just begun this quest of literary detective work. However, Basho's travel journals hinted that he was attentive to woodpecker behavior and to skylarks. What other birds were mentioned increasingly in his *haiku*? Basho's avian Zen theme has yet to be fully appreciated. The next chapter develops its implications.

24

Zen and the Daily-Life Incremental Training of Basho's Attention

Keep your mind clear like space, but let it function like the tip of a needle.

Zen Master Seung Sahn (1927–2004)[1]

Tell me to what you pay attention and I will tell you who you are.

Jose Ortega y Gasset (1883–1955)

By at least his third decade, Basho had become a student of Rinzai Zen. Although he never became an ordained monk, his close affinities with Zen were still evident in his clothing and tonsure. Traveling was risky in the late 1600s, and to be safer, he often wore the darker robes of a monk and shaved his head.[2, 3]

We don't know how many hours he spent in regular za-zen practice. On the other hand, he clearly exemplified a major emphasis in Zen training: the honing of keen attentional and intuitive skills during daily life practice (*shugyo*). These skills are on display as entries in Basho's five travel journals and in many hundreds of his *haiku*. The whole collection serves to document the remarkable content and scope of his informal training in the art of remaining acutely alert and aware, indoors and outdoors. Master Seung Sahn's epigraph points to the essence of these twin attentive functions. Clearly, Basho was a mindful observer, exercising both the requisite divergent and convergent creative problem-solving skills.

Birds play a special role in this regard.[4] The sightings and songs of birds provide excellent natural opportunities for training one's explicit powers of hearing and seeing. The data condensed in table 24.1 offer a novel way for a biographer to evaluate the literary, audiovisual, and implicit psychological consequences of this "avian presence."

Its columns provide a semiquantitative estimate of how birds influenced Basho's attentiveness in each of seven intervals during the last 33 years of his life. Birds not only flew into his journals, they would land easily on many lines of his *haiku*. Birds tell us who Basho *was*.

No such longitudinal textual analysis could have been possible without referring to the 1012 *haiku* and supporting details documented throughout Reichhold's excellent book. The table illustrates seven key points condensed from its pages:

- Basho started to write *haiku* when he was 18. During the next eighteen years—from 1662 through 1679—his 113 *haiku* cited birds only four times (three cuckoo, one gull). Obviously, during his first 35 years, Basho was not yet "a bird person."

Table 24.1
Bird Citations in Basho's *Haiku*

Phases of Development	Early Period	A Professional Poet	Retreating to Nature; a Spiritual Life	A Poet on Journeys	A 1500-Mile Journal to the North	At His Peak, Still Traveling	Maturing Further in Poetry and in Life
Interval [total years]	1662–1674 [13]	1675–1679 [5]	1680–1683 [4]	1684–1688 [5]	1689 [1]	1690–1691 [2]	1692–1694 [3]
Total *haiku* during each interval	53	60	73	304	138	147	237
Individual *haiku* (Reichhold's Sequences)	1–53	54–113	114–186	187–490	491–628	629–775	776–1012
Number of bird citations (% bold)	2	2	10 (**13.7%**)	90 (**9.9%**)	9 (**6.5%**)	21 (**14.3%**)	16 (**6.8%**)
Particular incidents of note	Studying Zen with Butcho sometime between 1673[a] and 1684		Takes his new pen name in 1681 from the plantain/banana tree, a gift from Rika.	"Old pond ..." in the spring of 1686 Temple, rain-drops, "original face" in the autumn of 1687	"The voice of a dove pierces my body." (summer of 1689, at Ogaki)		Bush warbler! Bush warbler! Water rail (two times)
Skylark[b] citations				2, summer of 1687 1, summer of 1688	1, spring/ summer 1689	1, summer of 1691	

[a] Reichhold dates to 1681 his starting to study Zen with Butcho (p. 409). Barnhill's date is 1682 (2005, p. 157).
[b] *Hibari (Alauda arvensis)* is the bird famous for the high-soaring male's melodious courting song.

- Soon, birds became "significant others." Notice the major change that took place during the next four years. It was sometime during this interval that he began his Rinzai Zen studies with Butcho (estimates vary). Then, from 1680 through 1683, he cited birds 10 times in the next 73 *haiku* (13.7 percent).

- During the next 5 year interval—from 1684 through 1688—Basho cited birds 30 times in 304 *haiku* (9.9 percent). Clearly, people who travel as he did, on foot outdoors, are more likely to see and hear more birds. We're interested in the cumulative effect that all these avian exposures would have on retraining and honing his *attentive* skills. Global and focal attention play a crucial role, poised at the foremost tip of all our subsequent mental processing. [MS: 13–20] So now, the central question becomes this: As each fresh event *captures* a traveling poet's attention, how do these *repeated* moments keep reshaping the plasticity of his brain, sensitizing his consciousness, and enriching his literary repertoire?

- Direct and indirect answers to this question require more conclusive facts than are now available. The two previous chapters have referred to three major stories. For example, one "story" in chapter 22 still needs the actual dates and places of the incidents relating to the "old pond" *haiku* in the spring of 1686. Having more details about a second story in chapters 22 and 23 could help motivate a search for further clues about "rain," "moss," and "moonlight" in the dialogues that Basho and Butcho had shared in different years.

In chapter 23, the third set of factual evidence would clarify the raindrops/"original face" event in the Buddhist temple during September, 1687. Events that night prompted two *haiku* and obviously made a deep impression on Basho.

- A much more subtle index of potential answers exists. We begin with the arrival of a very special *high-flying* bird, *Alauda arvensis*, the Eurasian skylark. Not until the summers of 1687 and 1688 did Basho begin to cite skylarks. The songs and flights of male skylarks are extraordinary. It was the skylark's song that motivated Vaughan Williams and Hoagy Carmichael to compose memorable musical scores.[5] If you hope to see this aptly named bird high aloft, you'll need to express the capacity to look *up* and *out* into the distance. [MS: 74–79] This means you'll be extending your neck, tilting your head way back, then gazing *far up* to focus your search on a faint speck far out in the distant sky. [ZBH: 97–151]

• Basho became alert to the fact that, for him, a bird song was a direct, penetrating, lived experience. For example, in 1689 he wrote "the voice of a dove pierces my body." In that same year, he specified the cutting word, *ya*, in *haiku* 509. This emphasized the distinctive, piercing archaic call of the crane. Only during his last three years were the unique liquid warbling notes of the *uguisu* also emphasized by *ya*[!]. Only then was the special *rap* of the water rail described as "door-knocking." These bird songs have a special impact. They are being *felt*, not just heard.

• Basho's *haiku* during his last five years cited the cuckoo 23 times. It was his all-time favorite. D. T. Suzuki comments on one reason why the cuckoo is a favorite of Japanese *haiku* poets: it sings in flight. Even though you may hear this song moving through the night air, you can't see the source.[6] Among Basho's second tier of (all-time) favorites, the next five birds also voiced prominent calls. This group included cranes (seven times), sparrows (six times), bush warblers (*uguisu*) (five times), plovers (four times), and pheasants (three times). No other nonhuman creatures, airborn or ground dwelling, attracted so much repeated attention in Basho's *haiku* as did birds. This need not come as a surprise. In the instantaneity of direct experience, bird sightings and bird songs tap instinctual levels, reminding us subconsciously of our deep roots in the natural world.

In Zen, enlightenment implies a "lightening up" of old rigid attitudes *and* behaviors. Not until you briefly let go of all the heaviness that had formerly preoccupied your mind and body do you deeply realize what it means to *really* "lighten up."

In his final years from 1690 to 1694, Basho employed one particular term to characterize his poetic technique. He called this theme "lightness" (*karumi*).[7, 8] He also applied this word to literature in general. *Karumi* meant a style of writing that was simplified, uncontrived, effortless. Returning to Blyth's literary formulation back in chapter 22, this theme of lightness would imply that the "doer" now performs any work with a "feather-light touch." No Self-conscious sense of accomplishment exists because the "deed" is done with no clinging attachments. Each event seems to happen by it-self out in the whole wide world. Free from the heavy bur-den of intrusive personal emotions and useless details, the liberated writer's perspective becomes more objective, flexi-ble, direct. For Basho, it was when he was in the outdoors that Nature's fresh stimuli became the meaningful "sensa-tions" that often culminated in his *haiku*. [ZB: 664–667]

Karumi is an elastic word (like *samadhi*, earlier, or *wu-wei*). Its lightness can have sensorimotor associations that feel in-ternalized or externalized. It can also have shallower (so-matic) or deeper (psychic) resonances. For example, after a state of internal absorption, the effort-free, brisk lightness of movement can keep flowing spontaneously for many hours. Embodied in this same sense of immediacy is a distinctive feeling of actual *physical* lightness. [ZB: 508–510] Had Basho recognized *karumi* as a term because he had also "*been* there"?

Alternatively, the impression in *kensho* is of being released from the heavy burden imposed by every last *psychic* form of bondage (Skt: *moksha*). [ZB: 536–539] *This* lightness is experienced in the depths of the psyche. Consciousness will finally arrive at its innate clarity, its freedom and competence, only when it has also dropped off all of its former overconditioned longings, loathings, and existential delusions. [ZB: 611–617]

When each of these several dynamic states of lightness impacts *directly*, it is readily apparent at anyone's first-person level of experience. It is tempting to consider that Basho so emphasized *karumi* after he had appreciated such similar qualities emerging in himself, qualities with which he could then identify with consciously. However, those sharp distinctions drawn above between the states of absorption and *kensho* were made for preliminary purposes of discussion. These differences become less obvious when we can read about them only at a third-person level, when all the old facts are not yet in, and when the few words now being examined are only in an English translation.

Can One or More Deep States of Awakening Enhance the Maturation of Affirmative Behavioral Traits?

In the course of "growing up," many normal adults also let go of their overemotionalized burdens. They "lighten up" and undergo beneficial "passages" to a more sane lifestyle. The term *passages* has been used to refer to these normal, ordinary psychophysiological transitions toward maturity that unfold over the years. [SI: 221–248; ZBH: 20–21, 91–92, 223–224 n. 6] Could relatively simple measures—those that cultivate one's focal and global attentiveness—exert a favorable incremental influence on the *normal* plasticity of the brain, further reinforcing one's *inherent* developmental patterns of psychological maturity?

It depends. Lesser epiphanies enable us to pause in natural outdoor settings. However, occasional minor quickenings are not central to the deeper issue now being examined. [ZB: 21] A large survey of mystical experiences in the general population showed that when such brief pauses were limited to *aesthetic* responses per se, they did *not* cause a substantial enduring change in the person's subsequent religious orientations or interpersonal relationships.[9] Then which people were transformed the most? These were individuals in a different category. They had undergone *repeated* experiences of *various* kinds, both aesthetic *and* religious. In the case of Basho the poet, these pages suggest that in his later life he experienced a variety of larger and smaller openings in his consciousness. They become a plausible explanation for why Basho *and his haiku* continued to show substantial maturation during his later forties.

Another significant research report was based on a longitudinal study of 31 male and female monastic trainees in the Soto Zen tradition.[10] Seven of these trainees had experienced the extraordinary state of *kensho* during their five-year period of residence. Serial psychological evaluations suggested that these brief states of awakening could have enabled them to mature more rapidly than did their cohorts and to have become more integrated and better-adjusted individuals.

Nielson and Kaszniak studied 16 meditators who represented different Buddhist traditions.[11] When these subjects had followed a regular meditative practice for more than 10 years, they appeared less attached to their physical symptoms. They also reported greater degrees of "emotional clarity." This important capacity for clarity was defined as "the ability to accurately discriminate among, and label, one's feeling states." Their greater degrees of emotional clarity correlated with both lower arousal ratings and lower skin conductance responses. These meditators were also more skillful in discerning which emotional valance was present even when it was hidden inside a masked picture taken from the International Affective Picture System. [ZBR: 245–

246] Good poets could become even better poets when they cultivate these qualities of subliminal discernment.

The Final Days; Autumn 1694

It is late September. Basho is ailing. He will compose only a few more *haiku* before he dies. He wrote this one as a greeting for the hostess who held a poetry party[12]:

> white chrysanthemums
> looked at closely
> no dust at all

Set against the pure white mound of a chrysanthemum, even a tiny speck of dust might stand out as an imperfection. Hamill[13] notes that such a dust mote might be Basho's oblique reference to another early incident in the Zen legends about Huineng.[14] Before this monk became the Sixth Zen Patriarch, he pointed out that no dust could alight on and mar the surface of a mirror (or of anything else) that from the outset had no permanent, independent existence. [ZB: 570–572] Multiple incidents in his daily life would remind Basho of earlier references in the classical Chinese and Japanese literature, including the old lore of Zen.

In Conclusion

Basho's poetry and journals during his last decade illustrate a keenly informed, highly developed approach to a mindful, incremental daily-life practice. This evidence suggests that as he began to see more deeply into the world of Nature, he then became aware of each of its diverse inhabitants—large, small, animate, inanimate, *and especially* those that were feathered. Moreover, that distinctive sound he had heard in the past—a frog plunging into water—was already becoming widely recognized as a unique sound-image. In the West, three centuries after that water-sound, do any of its distant ripples still play a subtle role in encouraging today's popular versions of three-line poetry?

Consider one recent example: This is a slim (1 cm), small (15 by 15 cm) booklet with only 60 pages. Yet even its succinct title, *Zen Birds*, speaks volumes. With a feather-light sense of touch, Vanessa Sorensen's wet brush skillfully portrays the tree swallow in soft watercolors.[15] Then, on the facing page, are her lines of verse. They exemplify *karumi*, the airborne essence of this bird:[16]

> How lucky the air
> To feel the graceful embrace
> Of the swallow's wing

Today, does it matter *where* certain details of the Basho stories occurred, *when* they happened, and *to whom*? Yes, facts matter. Given today's immediate global networking, we've been led to expect that when any revolutionary incident occurs now, it will quickly capture the world's attention and be (over)documented with detailed factual reporting. Clearly, most events relevant to these Basho narratives are now centuries old, beyond reach of such current expectations. Even so, this evidence hints that Basho seemed to have undergone *two* substantial openings, not one. Moreover, he was repeatedly focusing his attention on daily life events, actualizing countless present moments in memorable *haiku*.

Two substantial *haiku* and a journal entry date to that particular rainy night and furtive moon of late September 1687. The episode deeply penetrated him and appears to have opened into an alternate state somewhere near either absorption or *kensho*.[17] It did more than inspire *haiku*. It also left more actual first-person journal entries in its wake—waiting for us to clarify—than did that iconic frog-water sound, the echoes from which are still going on worldwide.

Meanwhile, this remarkable man clearly accomplished much during the final eight years of his short life. He became not just a revolutionary poet in his own day, and a saint to some, but an exemplar of Zen poetic sensibilities for all time.[18] If the ambiguities pointed to in these pages are to be viewed correctly, they will be seen as a plea for more facts, including dates, that can inform a more coherent narrative. And if these hints motivate others to clarify what *did* happen in Japan over three centuries ago, then they will have served their intended purpose.

25

A Story about Wild Birds, Transformed Attitudes, and a Supervisory Self

I hope you love birds too. It is economical. It saves going to heaven.

Emily Dickinson (1830–1886)

M.L.K. is a Doctor of Veterinary Medicine. When she was 34 years old, she experienced two unusual events. At that time, she was actively involved in a hazardous wildlife conservation project—saving the lives of wild parrots in Guatemala. Poachers were active; guerilla warfare existed in the community.

During the weeks leading up to the first event, she felt happy and joyful in general, fully engaged in her work, and in a heightened mode of arousal. She arose one morning before sunrise during the rainy season. She and her familiar local guide had planned to take a five-mile hike. They were searching for parrot nests in one of the rare forest tracts that had not yet been burned off to clear the land for cattle. While walking around this ranch, during casual conversation, the guide mentioned why he deeply loved Mary and Jesus as more than mere symbols, and also loved the Catholic Church in general. She was reluctant to accept his explanation because at the time she felt that she was being proselytized.

They then reached a place where tall trees, standing a hundred yards away, were being illuminated from behind by the newly risen sun. In the humidity, the moist tree leaves were releasing a mist that heightened the effect of this *contre-jour* illumination by the light of the sun.

A group of parrots were perched higher up in this canopy. All at once, the whole flock rose into the air. They flew from right to left, crossing over in front of her, emitting loud squawking calls. Awed by the beauty of the entire setting, surprised by the flight of the birds, and overcome by a sense of tenderness, her legs gave way. She fell to her knees on the ground. There, she also felt embarrassed. Never before had she been so dissolved by such a combination of visual and auditory stimuli.

"This event *softened* me. My attitude underwent a deep transformation. From that exact moment on, my experience became one of greater openness whenever my friend or anyone else tried to proselytize me. Now, I understood—when any person spoke about the Catholic Church using terms like Mary and Jesus, this was *their way* of simply saying trees-birds-sun. Now the Church had become another way of talking about birds and trees and a world filled with beauty."

Indeed, after this episode, she became much more active in the local Catholic Church. (Eventually she would enroll in a nonsectarian divinity school in order to "learn the language of the human heart." This would then lead to her becoming an ordained Unitarian-Universalist Minister.)

The second transforming event occurred six months later. She was now much more aware that large numbers of adult parrots were being poached. This meant that many of their abandoned chicks were left to starve in their nests. Yet her time to work in Guatemala was about to expire. With feelings of great anxiety, she left instructions on how to care for these orphans, then rushed off to the airport to fly back to her home in the United States.

Once on board the airplane, she found that its departure was delayed. Left to herself in a half-full cabin, she took her seat by the window on the right side. There, she started to read a veterinary journal in order to pass the time. Suddenly, in the quiet, she heard her own first name spoken. A voice repeated this first name three times during the next two minutes. Each time, she looked around but saw no speaker in the cabin. Once, she even arose and searched the whole inside of the plane including the bathroom and the cockpit. She found no one who could have spoken her name.*

At this point, she sat down, then heard the voice call her name the fourth time. She heard herself say, internally: "O.K., I'm listening." The voice then said three words: "Move to Guatemala." This same message was repeated two more times during the next minute or so. She responded briefly to the voice with internal words of her own, resisting its advice. For example: "I can't. It's too dangerous. It will damage my career ..."

Finally surrendering, she said "O.K., I'll go." As soon as this big decision resolved itself, she burst into tears of relief and gratitude. During the next hour, her emotions evolved

---

* She had never before, nor since, experienced any hallucinations. She was a nonmeditator and had grown up "unchurched." The voice was more androgenous than male. It had no other distinguishing features. Her seat on the right side of the cabin might suggest that the voice could have arisen more off to her left. However, at present, she recalls no lateralized source.

into this deep, clearest realization: *Love* is the most important thing in the world—*love* for both birds and people.

About six months later, she was finally able to move back to Guatemala. There, starting in the summer of 1992, she felt euphoric about this conservation project for her first six months. She remained there for three more years in spite of increasingly difficult living and working conditions. In retrospect, she continues to regard as a very special "gift" the deep clarity of that life-changing event that happened spontaneously while she was waiting for the airplane to take off.

Commentary

This is a true story about the internal origins of speech in a nonmeditator. How is it related to the precipitating causes of other openings that meditative training is intended to cultivate?

   • A common theme emerges in four recent accounts. Readers may be reminded that a propitious setting had triggered each visual shift toward an alternate state of consciousness. Earlier chapters emphasized the *allo*centric mode of such a shift and discussed the implications it has for our understanding of psychophysiology.

Chapter 23 described one such an illuminating situation. Basho is at the temple, *gazing upward*, finally seeing *moonlight* glisten off the foliage of wet tree leaves. In the present chapter, M.L.K. is also *looking up* toward the tops of tall trees. In the early morning mist, she is seeing moist leaves illuminated from behind by the *sun*light. In these two special settings, an unusual quality and quantity of light is coming from *out there* and is entering the person's visual fields (superior > inferior fields).

The experiences of two other subjects, K.K. and O.R., were described in detail in 2014. [ZBH: 99–123] During times of stress, each woman was also looking *up* and *out* into the distance. The distant trees in O.R.'s case were also being illuminated by moonlight. Research has shown that the presence of trees in an outdoor setting confers many intangible benefits (appendix A). [ZBH: 33–48, 186–189] Subtle arboreal benefits also enter into the narrative that recounted this author's recent Nature walk through a woodland sanctuary (chapter 10).

About Silent Speech (Word Thoughts)

- Voices that articulate advice tend to arise from an overactive brain. Often the person is in an overstimulated mental state, sometimes troubled by conflicts that seem intractable.

Augustine of Hippo (354–430) was in such a state of turmoil. He was struggling both with his conscience and with the fixed belief that he himself lacked enough willpower to restrain his ingrained lusts for women and wine. At this crucial point, he heard a voice that came from a neighboring garden. It said, "Take up and read. Take up and read." Fortunately, a Bible was nearby. Opening it, the first passage he came upon was Paul's Epistle to the Romans, 13, verses 13–14. There he discovered the antidote for his longstanding lusts of the flesh. The solution was clear—become one with "the Lord Jesus Christ."

As David Brooks points out in his book about character,[1] Augustine had already embarked on a Christian path. This was a further "elevation scene." It was not the kind of full-fledged 100 percent conversion event of a nonbeliever that Saul of Tarsus had undergone on the road to Damascus. It was transformative, however, to the degree that the voice

Augustine heard finally set him on the path of submission to what his belief system informed him was a higher power.

For a confirmed Christian believer of that era, a transformation of Self was not up to him. Instead, he would have to renounce "the whole ethos of Self-cultivation." What Augustine had to accept was the doctrine that he could only be saved by the amazing Grace that came down from the highest ultimate source.

Parenthetically, from a Zen Buddhist perspective, awakening arrives from no such lofty transcendent source. It arrives intuitively when men and women follow the ennobling eightfold Path on *this* Earth (chapter 2). [ZB: 144, 356] In this belief system, the immanence of Nature's world becomes a sturdy altar toward which one tends to bow in gratitude and worship. [ZB: 554–556]

> • The ordinary word thoughts that we hear during inner speech are a different matter. Normally, we *know* when we are "talking *to* ourselves." [MS: 146–150] We're certain that *we* are the agency uttering the word-thoughts of this *inner speech*, not some outside, ill-defined source (the kind that seems to enter from the outside during an auditory hallucination) (chapter 10). While M.L.K. was waiting on the airplane, she also knew that those later words she "heard" when she resisted that other seemingly "external" voice were *her* own internal words, in *her* own voice.

It is estimated that we spend around one fourth of our conscious waking life in various silent forms of inner speech.[2] Neuroimaging studies suggest that the more conversational (dialogic) forms of our inner speech scenarios activate the superior temporal gyrus, precuneus, and posterior cingulate gyrus bilaterally as well as the left inferior and left medial frontal gyrus.[3] Of course, the normal ruminations of this discursive speech can help us or harm us. Much depends on how we interpret the source of the message and what it means. Do we accept such words as real and true? Or are they just the same ordinary word-thoughts that originate in us?

- What are some benefits of generating fewer word thoughts? Regular, long-term Zen meditative training enables one's excessive, Self-driven, maladaptive varieties of inner speech *to drop off*. This liberated field of consciousness becomes free from its own inner noise. Now it can perceive the world's signals with greater clarity. [ZBH: 148, 155–158] But notice: this improvement in the signal/noise ratio doesn't mean that *all* mind wandering is deleted. The resulting clarity just leaves more open room in which useful logical thinking can occur. Moreover, constructive forms of mind wandering can develop some divergent associations that happen to be more creative (chapter 1). [ZBH: 78–80]

A 50-year review of the literature on this "little voice inside my head" cites 190 references. It closes with a sobering note: the issues addressed are still "far from resolved."[4]

- The emphasis during Zen training remains on direct perceptions and immediate actions, not on intricate concepts. Abstract concepts are more likely to be attached to sticky layers of complex meanings. Even though these meanings do *not* necessarily depend on words, the linguistic aspects of meaning still exert an entangling subterranean influence. [SI: 130–143]

This makes word concepts very difficult to investigate. Undaunted, Wilson-Mendenhall et al.[5] decided to restrict their study to abstract concepts that were intrinsic to actual *visual scenes*. Under these conditions, the separate meanings of *convince* and *arithmetic* were grounded in "distributed neural patterns that reflect their semantic content." Their experimental subjects turned out to represent convince and arithmetic in such widely-distributed sites as the posterior cingulate, precuneus, right parahippocampus, and lingual gyrus bilaterally (figures 6.1, 6.2, and 15.1).

- So, our *associations* are the basis of the meanings that we infuse into *everything*, tangible and intangible. Meanings emerge only when multiple interactions link multiple regions. [SI: 143–152]

Your own associations are what helped to bring meaningful coherence to the eclectic range of topics on these pages. [ZBR: 229–232] Your vast array of white matter pathways serves as the "wiring" for these connecting links. You can see full-color representations of these networking pathways. They're parts of what has now been called "the Human Connectome"[6] (appendix D).

Keep elevating your gaze and reminding yourself to cultivate your associations on these networks with care. ...

In Closing

A Living Zen heightens our open-minded conscious sense of deep *appreciation* for the entire natural world. Living Zen also enters subliminally through open eyes and ears into an open-hearted processing of whatever arises in this present moment. We bow in gratitude, remindful of the countless gifts that all other beings contribute to our living on this incredible, fragile planet in this awe-inspiring Universe.

The lines in Ryokan's final poem[1] remind all Earthlings how much we share the natural outdoor world with other forms of life. In parting, he asks:

> My legacy, what will it be?

He answers:

> The flowers of springtime,
> A cuckoo in the summer,
> The scarlet leaves of fall.

Parting takes on special resonances in Japan. Buddhists might say "*Odaiji ni*" to the person who is about to depart. The phrase implies "Please keep directing your attention toward the 'Great Matter.'" This means toward the Ultimate *Eternal* Reality, toward the Big Picture, way past any notions clinging to one's own living or dying. By way of reassurance to those who enjoy reading, such a journey still includes reading books about Zen, especially where they encourage the perspective that direct experience is more useful than words.[2]

Appendix A: Back to Nature: Pausing in Awe

> The clearest way into the Universe is through a forest wilderness.
>
> John Muir (1838–1914)

> Keep a green tree in your heart, and perhaps the singing bird will come.
>
> Old Chinese Proverb

> How many times must a man look up before he can see the sky?
>
> Bob Dylan, *Blowin' in the Wind*

The world of Nature has evoked awe in many human beings and inspired them to follow a spiritual Path. [ZB: 664–667; ZBH: 186–189] Piff and colleagues became interested in awe.[1] Their series of behavioral experiments confirmed that awe can diminish Self-centeredness and increase other-directed (prosocial) forms of behavior. When subjects were tested for their generosity in a money-based economic game, those who already tended to experience awe gave most generously, even more so than did those who were disposed to compassion and to all the other prosocial emotions. Subsequent experiments during which awe was induced increased the likelihood that these subjects would then make ethical decisions, be more generous, and feel more inclined toward prosocial values in general.

The researchers induced awe in one experiment with the aid of a unique natural setting. It was full of tall trees. The subjects stood for one minute in this special grove of towering eucalyptus trees. This brief, actual immersive experience outdoors enhanced their subsequent prosocial helping behavior and decreased their sense of entitlements in comparison with controls.

When subjects witness the almost cathedral-like atmosphere of towering green trees overhead, they are receiving several benefits of visual messages throughout both their superior and inferior visual fields (chapter 10). The authors propose that the "feelings of a small self" are at the core of these beneficial effects that awe has on prosocial behavior. Such a diminution of one's own Self-centeredness could seem to enhance the likelihood of one's becoming interested in other persons and more aware of broader social contexts.

In a related study, college students were surveyed online for various (self-reported) positive emotions.[2] Their positive emotion of awe was the strongest predictor for *low* levels of the harmful proinflammatory molecule interleukin-6.

It is plausible to propose that such "feelings of a small self" could also have testable neuroimaging correlates. However, it becomes open to question just how far today's so-called "virtual reality" electronic equipment could provide an accurate approximate equivalent of the "*real* thing."[3] Deep, *authentic* awe is a sublime event that is usually experienced in the real outdoors. The occasion is often spontaneous, not contrived. Moreover, the infinitesimal witness is immersed in a night sky that knows no ceiling, or is in some other Grand Canyon-like natural setting that expands time and space far beyond one's ordinary boundaries of mere wonderment. The current overuse of the word "awesome" suggests how far researchers still have to go in order to understand real, authentic, deep, natural awe, all of its psychophysiological correlates, and each of the mechanisms involved in its subsequent triggering effects.[4]

Appendix B: Reminders: The Crucial Role of Inhibitory Neurons and Messenger Molecules in Attentional Processing

> Science prospers exactly in proportion as it is religious; and religion flourishes in exact proportion to the scientific depth and firmness of its bases.
>
> Thomas H. Huxley (1825–1895)

Our brains have been overconditioned for many years before we started to meditate. Much of the Zen meditative approach is then *re*training. This is basically psychophysiological, not philosophical. Consider the simplest system: two nerve cells joined by a synapse. Let neuron A release an *excitatory* transmitter out at its nerve endings. When this molecule stimulates its receptors on the next neuron B, they *increase* that cell's firing rate. But let another neuron, C, release an *inhibitory* transmitter from its nerve endings. Now, the inhibitory receptors on the next cell will *decrease* the firing rate of this neuron D.

On the other hand, suppose this same inhibitory neuron, C, had also been holding in check the firing of a different excitatory neuron, E. When this same inhibitory neuron C *itself* becomes inhibited, then neuron E will be *released* from this prior inhibition. Now neuron E fires much faster. This example illustrates the vital role that *disinhibition* plays in a three nerve cell system.

Imagine the possibilities when our *whole brain* organizes galaxies of individual neurons into different nerve assemblies! Up in the cerebral cortex, these assemblies take the form of columns of nerve cells packed into layers. This cortex is then folded into hills called gyri and valleys called sulci. Farther down in the subcortex and brainstem, the larger assemblies take the forms called nuclei. The hypothalamus, pulvinar, and amygdala are examples of these deeper nuclear structures.

In this much larger context, these cortical and subcortical modules are linked in networks that are constantly *interacting*. These interactions—some feeding forward, others backward—blend countless activations with *de*activations. When does each cluster of nerve cells now become most excitable? Only during propitious *oscillating* sequences. Big clusters of neurons fire only when multiple impulses arrive on their dendrites and cell bodies at precisely the right time, in precisely the right numbers, and on precisely the most appropriate receptors. Indeed, these responses become most efficient when just the right oscillations take place that blend their collective excitations, inhibitions, and disinhibitions. [ZB: 147–157]

Current studies of visual attention[1] and of auditory attention[2] explain why such *synchronous oscillations* play a crucial governing role in shaping our attentional processing and our resulting behaviors. Any two rhythmically interactive neuronal assemblies communicate most effectively when the peaks and valleys of their individual oscillatory phases become closely *aligned*. This alignment ensures that when bursts of impulses are sent from the first group, each volley arrives inside a very narrow open "window of opportunity." In fast-firing systems, these "windows" are reduced to mere slits that open for very few milliseconds. It is incredible that anything like this actually works. Yet the hummingbird's wings beat 70 times a second, even when it is just hovering.

Zen lore emphasizes that awakened states can be triggered unexpectedly by stimuli that strike at random (chapters 11–13). This reflects the fact that the multiple components of our brain networks exist in dynamic, nonlinear, interactive configurations. How does the neuroanatomy in chapter 6 translate into psychophysiology?

Logothetis' review of primate triggering[3] describes one pertinent example. This cascade of events begins in the CA 3–CA 1 hippocampal circuit (figure 6.1). These sequences can facilitate the optimal transfer and consolidation of declarative memory messages between the hippocampus and the neocortex. When CA 3 neurons fire in bursts, they depolarize CA 1 dendrites. The resulting fast activities take the form of various *sharp ripple oscillations* (100–200 cps). These "ripples" represent *release* phenomena (disinhibition). Ripples become prominent *after* theta rhythms from the medial septal region have first been suppressed.

The onset of sharp-wave rippling oscillations correlates with two robust contrasting events: strong activations of both the neocortex and the limbic cortex; down-regulations in the thalamus, basal ganglia, cerebellum, and brainstem regions.

What about the *reticular nucleus*? [SI: 92–93; ZBH: 118–119] Do the latest models simplify what we had previously known about its major *inhibitory* role in the actual operations of human thalamo ↔ cortical networks? Not yet. Current models of this thin inhibitory nucleus still leave much for future research to discover in humans and to simplify.[4] This is especially true of the vital sleep-waking transitions of normal human subjects who are also meditators.[5,6,7]

Meanwhile, discrete sectors of the reticular nucleus can be stimulated optically in mice.[8] Tonic stimulation for 30 seconds promptly suppresses the activity of the underlying thalamus. Within 20 ms, delta slow wave activity appears up in the corresponding region of the cerebral cortex. As the resulting slow-wave patterns oscillate within this reticulo-thalamo-cortical circuitry the mice show decreased behavioral arousal.

Mice are rapidly aroused from their normal slow wave (non-REM) sleep as soon as the firing of GABA neurons down in their *lateral* hypothalamus powerfully inhibits other GABA neurons up in their reticular nucleus.[9] This disinhibitory release phenomenon could provide a potential explanation for certain early arousal events during internal absorption. [ZB: 478–481] (chapter 16)

Meanwhile, models based on the results of magnetoencephalography (MEG) confirm that attention plays a pivotal role in sharpening the clarity of our sensory processing (appendix C). Moreover, this heightened effect of attention is being linked to the actions of *inhibitory* interneurons. The brain's key *inhibitory* transmitter is GABA (gamma-aminobutyric acid). An astonishing fact is that our brain needs only one short enzyme sequence to make this essential inhibitory amino acid transmitter. What is its precursor molecule? Glutamate, our key *excitatory* amino acid transmitter. So, in this one synthetic step—glutamate → GABA—we generate the key inhibitory molecule that can save our brain from being stimulated into epileptic seizure activity. [ZB: 208–209]

Yet, GABA can also function in other disinhibitory roles that translate into net *excitatory* functions.[10] Many GABA neurons in the basal forebrain promote fast cortical oscillations. These fast-firing nerve cells (20–60 cps) ultimately enhance neocortical gamma oscillations. They are implicated in active states of wakefulness and REM sleep.

Let one cholinergic nerve cell release its transmitter, acetylcholine, on the *nicotinic* receptors of a second cell, and the resulting cholinergic excitatory effects are immediate. Yet, suppose this same transmitter molecule were to be released on the *muscarinic* receptors of a third nerve cell. Although these excitatory responses begin slowly, they continue for many seconds. [ZB: 164–169] In contrast, our biogenic amines act as *modulators*. The modulating functions of dopamine, of norepinephrine, and of serotonin systems add subtle secondary interactive layers to those primary functions of excitation that glutamate and acetylcholine confer and to those major primary higher functions of inhibition that are chiefly conferred by GABA. [ZB: 197–208]

What makes certain triggering stimuli especially salient (chapters 11, 21, 23, 25)? It makes a difference *which* hemisphere is being activated. The *right* hemisphere is the more sensitive when stimulus cues enter the opposite (left) visual field. This is called "the left visual field advantage."[11] Our attention is especially captured when a stimulus is unexpected, when it is novel, or when an expected event is suddenly omitted (a torchlight extinguished at night).

Acetylcholinergic nerve cells of the basal forebrain are responsible for much of the direct attentive salience attached to these alerting stimuli.[12] [ZB: 167] These salient enhancements with novelty are linked not only to the cholinergic neurons of the substantia innominata and diagonal band.[13] Their alerting circuits can also include the central nucleus of the amygdala and the substantia nigra compacta.[14]

Last, but not least to keep in mind is nitric oxide (NO•) (chapter 17). This gaseous signaling molecule plays pro and con roles that have yet to be fully explored in the human brain. Many different neurons can also release NO• as soon as their excitatory glutamate and aspartate receptors are activated. Acetylcholine nerve cells are among these many other neurons that are involved not only in attention but are also poised to make NO• and to co-release it. [ZB: 412] Anything that involves attention and its allied states of awareness remains relevant to Zen.

Appendix C: Magnetoencephalography

The visual system reacts to stimuli very fast, with many brain areas activated within 100 ms. Using MEG, we show that the visual system separates different facial expressions well within 100 ms after image onset. This separation is processed differently depending on where in the visual field the stimulus is presented.

L. Liu and A. Ionannides[1]

I was skeptical when I first saw the shielded magnetoencephalography room at MIT. Two decades ago, it seemed a farfetched notion that nerve cells could generate enough energy to induce a *magnetic* field, let alone one that could then be usefully detected by equipment outside one's head.

I was wrong. Technical advances in magnetometer hardware, software, and computation greatly improved the early MEG. Responses from both the cortex and subcortical structures are being localized with a precision not possible by EEG. [MS: 106–108] The fact that MEG activities correlate directly with synaptic neural activity, within only a few milliseconds, distinguishes it from fMRI. fMRI can have a lag of five seconds or so in responding to changes in signals that are *b*lood *o*xygenation *l*evel *d*ependent (BOLD).

MEG laboratories are expensive, yet have other advantages. Meditators can sit, as they usually do when meditating. The silence is a welcome relief from the hammering and pounding of the fMRI suite.

The following sample of uses for MEG illustrates some of its special assets.

- Liu and Ioannides monitored seven right-handed men with MEG while they were viewing happy, fearful, or neutral faces.[2] When these faces were presented to the *center* of their visual field, the subjects quickly separated these three different emotions. The relevant attention processing signal appeared first in their *right* superior temporal sulcus (STS) at 35–48 milliseconds (chapter 15). It arrived next in the right amygdala (at 57–64 milliseconds), and then up in the medial prefrontal cortex (at 83–96 milliseconds). Suppose these faces were presented farther out in the *peripheral* quadrants. Now the signal peaks were separated first in whichever amygdala was on that same side as the facial stimulus *and* in the STS of the opposite side. The authors interpreted this amygdala's rapid response to the peripheral fields as an appropriate visual alerting step by subcortical centers. In contrast, the role of the STS was considered to facilitate a more cognitive appraisal of the situation by helping to link cortical sites together.

During these behavioral responses, the fearful faces were recognized most accurately when they were presented in the center of the visual field. Notably thereafter, the next best fearful face recognitions occurred during presentations in the *right upper* field. These were followed by presentations in the left *upper* field ($p < 0.003$), and then in the right lower field ($p < 0.05$). [ZBH: 111–120] The data suggested a slight *upper* field advantage.

- Luo et al. conducted a classic study of how first we respond to an emotion when we see it expressed on another person's face.[3] Their normal subjects glimpsed pictures of faces for a mere 0.3 second (300 milliseconds). These faces expressed either fear or anger or were emotionally neutral. A subcortical pathway responded quickly to the faces showing fear. These gamma responses in the amygdala arrived as early as 25 milliseconds, especially on the right side. They then peaked at 235 milliseconds and lasted for a total of 280 milliseconds. The response to anger, on the other hand, began much later in the amygdala (at 155 milliseconds), although it peaked at 215 milliseconds. However, it lasted for only 110 milliseconds and was more prominent on the left side. Information of this kind at this depth is not obtained by standard fMRI or EEG techniques.

• In the study by Kveraga et al., when a visual stimulus was correctly identified, MEG indicated that the left orbital frontal cortex was activated only 130 milliseconds after the stimulus was presented.[4] This early frontal activation could enable an object to then be identified more accurately by the subsequent (top-down) relay that facilitated the functions of the infero-temporal cortex. Activations did *not* occur at this site until 50–85 milliseconds later.

• Ioannides et al. studied normal sleep using high-resolution MEG techniques.[5] They found that gamma activities increased in a large region (5 cubic centimeters) in the *left* dorsal medial prefrontal cortex. Gamma activities increased in this core region during REM sleep to levels even higher than during active wakefulness.

• Lou and colleagues[6] estimated that the spatial resolution of MEG was at least 10 millimeters in the cortex and 20 millimeters in the thalamus. Their maximum activations were present in the lower gamma frequencies between 30 and 45 cps. These gamma oscillations were strongly enhanced during an autobiographic memory retrieval task. During this task, their subjects recalled judgments they had made earlier about adverbs that were presented visually. Granger analytical tests identified the time sequences of the resulting signal peaks. During the remindful activities of this particular memory retrieval network, the *pulvinar of the thalamus* was connected with the *medial prefrontal/anterior cingulate cortex and the posterior cingulate cortex*. The following simple diagram condenses the nature of this important medial Self–other continuum (chapter 5). It emphasizes the important subcortical role that the pulvinar plays in this autobiographical memory task. [SI: 87–90]

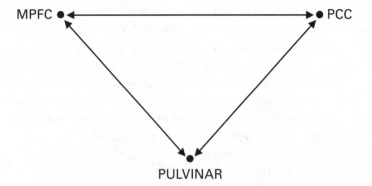

MPFC ●◄─────────────────────► ● PCC

PULVINAR

- Kerr et al. investigated the basic 7 to 14 cps alpha MEG rhythms that occur in the early primary sensory cortex.[7] They found that the eight-week course of mindfulness-based meditative training enhanced this basic alpha modulation of the primary sensory cortex (S1) in response to a cue.

- Styliadis et al. used a wide MEG frequency band of 2–30 cps activity to examine 10 adult subjects.[8] The differences in emotional valence between pleasant and unpleasant pictures appeared as early as 25 milliseconds. The differences between high and low arousal conditions began later, around 250 milliseconds. The *left* centromedial amygdala processed highly arousing stimuli that were also pleasant. In contrast, stimuli that had a negative emotional valence correlated more with responses in the *right* basolateral amygdala. [SI: 176–177; 251–252]

MEG can outstrip the mobility of our conscious thought, offering a millisecond glimpse of the sequence of operations inside of our "black box."

Appendix D: Diffusion-Weighted Imaging

> Following the cascade of changes in diffusivity may be a means to better understand the different processes of tissue destruction repair mechanisms which can take place in several neuropathological processes.
>
> R. Bammer[1]

A full-color illustration that samples some of our "100,000 miles" of white matter fiber pathways is now available in your public library. It's in "Secrets of the Brain," by C. Zimmer and R. Clark, in *National Geographic* 2014; 225(2): 31–57. These various highways and byways can be demonstrated using the techniques of diffusion-weighted imaging.

Diffusion-weighed imaging (DWI) is a variation on the theme of magnetic resonance imaging (MRI). It exposes the brain to a shifting magnetic field. The resulting slight displacements of water protons create measurable shifts in the ways that these water molecules then diffuse through the microarchitecture of white matter. [ZBH: 160–161, 167, 241–242 n. 27] Viallon and colleagues' 2015 article is recommended as a review of different magnetic resonance imaging techniques.[2] Its diffusion weighted methods distinguish among (1) measuring the *extent* of water movement without regard to its direction (DWI), (2) evaluating differences in the *direction* of water movement to derive information about brain architecture (diffusion tensor imaging, DTI), and (3) assessing the *heterogeneity* of a tissue using diffusion kurtosis imaging—(DKI).

How do major alternate states of consciousness evolve on the meditative Path? This section has a major purpose: to increase the general reader's appreciation, and that of the neuroscience community, for the potential roles that DWI could play in studying this question. This approach follows the lead taken in applying DWI to the management of neurological patients with acute stroke syndromes (chapter 9). Other recent articles indicate that DWI can also be used to follow long-term changes in brain tissue. Similar DWI applications could include, for example, its usefulness in following subtle longitudinal brain changes that correlate with the rare ongoing stage called the Exceptional *Stage* of Ongoing Enlightened Traits (chapter 2).

In the same year as Bammer's early review (2003), another review indicated that DWI could localize as many as 14 thalamic nuclei in the living human brain! [ZBR: 169] In keeping with the ways this series of books has been oriented toward joint thalamic and cortical functions, this latest sampling of recent DWI articles begins with the study by O'Muircheartaigh et al.[3] These authors combined the architectural contributions of DWI with the resting-state capabilities of fMRI in 102 adult subjects. The resulting novel approach was described as "diffusion tractography-defined structural and functional connectivity." The remarkable full-color images from this extensive study illustrate a subset of seven types of networkings. They demonstrate that prominent *thalamo*temporal and *thalamo*frontal connections exist that link different regions of the so-called default-mode network. They include one subset of connections that exemplify the loops that interconnect our thalamus, cortex, and basal ganglia. These looping circuits are crucial. They are a major source for the incremental brain changes that allow us to *revise and retrain our behavior*. [MS: 133–139]

Question: Given the associations noted in chapter 9 among the onset of transient global amnesia, straining on a closed glottis, active physical exercise, and emotion, could microemboli be dislodged from the systemic circulation and then cause microembolic occlusion, *selectively* in the hippocampus? No. When microemboli from the systemic circulation enter the major blood vessels in the brain, they show no *selective* affinity for the hippocampus.[4] (Micro-occlusions do occur, however, in borderland zones that are relatively underperfused.)

When the anatomical connectivity of gray matter voxels (as determined by DWI) is combined with fMRI data, there emerges—for certain regions—a precise and fine-grained relationship between their anatomical connectivity and their function.[5] For example, with regard to how we first perceive certain items: (1) our perception of *faces* is referable to the fusiform face area, the occipital face area, and the superior temporal sulcus; (2) our perception of *scenes* is referable to the parahippocampal place area, the transverse occipital sulcus, and the retrosplenial cortex; and (3) our perception of *objects* is accomplished by the lateral occipital cortex and the posterior fusiform sulcus (figure 15.1).

Damage to certain thalamic nuclei interferes with *transcortical* interactions. Does this add to the short-term memory deficits of those patients who also have hippocampal dysfunctions?[6] DWI tractography identified different thalamic subdivisions in patients who also had a chronic epileptic focus located in the medial temporal lobe. Again, when combined with functional MRI, the connectivity data suggested that the patients' *short*-term memory performance hinged on those functional circuits that connected their thalamus with their prefrontal cortex. However, their *long*-term memory performance correlated with functional connections linking the thalamus with their *temporal* lobe (chapters 6, 7). Patients with hippocampal atrophy also showed some atrophy of their pulvinar nucleus on that same side.

Studies of Emotion

It's been clear that we dampen emotions using connections among the medial prefrontal cortex, the basal lateral amygdala, and the central nucleus of the amygdala. [SI: 234] Diffusion-weighted images now enable the technique called *probabilistic tractography* to further define pathways connecting other frontal lobe regions with the amygdala.[7] These "top-down" circuits include the dorsal lateral, dorsal medial, and ventral medial PFC as well as the orbitofrontal cortex.

In this same study of emotion, after 310 volunteers took a trait anxiety inventory test, they were then preselected into two groups of female subjects. One group was the low-anxiety group; the second group was the high-anxiety group. The values for fractional anisotropy (FA) were taken as an index of their white matter microstructure.

The subjects who were *low* in trait anxiety showed stronger *right* hemispheric connectivity between their amygdala as a whole and both their ventral medial prefrontal cortex *and* orbital frontal cortex. In contrast, those subjects who used *reappraisal* to regulate their emotion in a more sustained cognitive manner showed stronger connectivity between their amygdala and all regions of their *left* hemisphere. These regions on the left included the ventromedial prefrontal cortex, the orbital frontal cortex, the dorsal medial prefrontal cortex, and the dorsal lateral prefrontal cortex.

An especially strong connectivity was evident through the *uncinate fasciculus*. Up in front, this white matter tract links the ventromedial prefrontal cortex with the orbital prefrontal cortex. [ZBH: 167–168] These increases in connectivity were not found in those other white matter pathways that serviced the dorsal lateral or dorsal medial prefrontal cortex. The data support the earlier concepts of "top-down" emotional regulation. [SI: 228–244; ZBH: 179–182] They also help distinguish the neuroanatomical correlates underlying our anxious personality traits from those reappraisal skills that we can use to regulate our emotions.

A spreading depolarization occurs during episodes of migraine and following vascular occlusion. Can this depolarization be detected using diffusion weighted imaging techniques? The studies[8] were conducted in rats. However, the results are relevant to transient global amnesia, to the potential usages of DWI in major alternate states of consciousness, and to those many persons (including neurologists) who also happen to be susceptible to migraine. [ZBR: 306–312] In brief, these experiments showed that the diffusion-weighted multi-spin-echo technique had the potential to provide, simultaneously, both cellular *and* hemodynamic information.

When normal subjects respond to shifting color patterns, their DWI signals change farther along in the visual *association* cortex (V4), not just back in the primary visual cortex (V1).[9] Moreover, the DWI signals at V 1 and V 4 peak from 1 to 3 seconds *before* the BOLD fMRI signals reach their comparable local peaks. These earlier DWI responses hint that diffusion imaging is especially sensitive to particular physiological events that develop across successive synapses. This recent finding further supports the possibility that suitably modified DWI applications (as, for example, was accomplished using high angular resolution—"HARDI"— techniques)[10] could evolve in the future to become key components of a more sophisticated neuroimaging panel. In this manner, DWI could help clarify which neural mechanisms underlie our major awakenings and related states of consciousness. [ZB: tables 10 and 11, 302–303]

New techniques have emerged that can increase the high spatial resolution of DWI.[11] Another consideration is the substantial resolving power achieved using 7 tesla MRI as was discussed in the human entorhinal cortex (chapter 6, n. 11, n. 12). These developments imply that a working hypothesis of lingering residua in such states as *kensho-satori* and internal absorption can be challenged by applying suitable DWI and structural techniques of the highest resolution at the most appropriate times.

The ancient Zen teachings, in this so newly promising a field, call out for only the most exacting clinical judgments of claims to be "enlightened," and for only the most sophisticated techniques that science and technology can bring to such a major research endeavor. Rigorous peer review insists on further refinements of study design that are orders of magnitude beyond the primitive attempts ventured thus far.

A Notable Nucleus

What else do we need to restrain our maladaptive behavior? (Skt: *Shila*) Civilizations rise or fall depending on how they answer this crucial question. Our subthalamic nucleus plays a pivotal role during an outright suppression of responses. [MS: 136] The subthalamic nucleus is also activated during more deliberate slowings or switchings. What else is known about this important subthalamic system? fMRI monitored the performance of 16 adult subjects while they performed a stop-signal task.[12] Using DWI, it was shown that those subjects who were most efficient in inhibiting their responses also showed greater white matter structural connectivity linking their prefrontal cortex with their subthalamic nucleus.

One model best explained these inhibitory results. It suggested that the subjects' inferior frontal gyrus would increase the excitatory influence that their presupplementary motor area (pre-SMA) had on this subthalamic nucleus. The results could enable messages from the subthalamic nucleus to then speed up through the thalamus to inhibit the motor cortex. The data also suggested that reciprocal connections between this inferior frontal gyrus and the presupplementary motor area could allow each region to modulate the other. Both sets of data supported the concept that the subthalamic nucleus normally acts as a "brake" on our behavior. However, it won't suppress motor output until it is informed from elsewhere whether the most appropriate response under the circumstances is to prompt action or inhibition.

Appendix E: Some Newer Methods of fMRI Analysis

> It is in my judgment the greatest office of natural science (and one which as yet is only begun to be discharged) to explain man to himself. The knowledge of all the laws of nature,—how many wild errors—political, philosophical, theological, has it not already corrected.
>
> Ralph Waldo Emerson (1803–1882)[1]

We need to remind ourselves: color pictures that represent a neuroimaged brain are vast oversimplifications. Today's "natural science" (itself only recently deployed) glosses over not only the intricate hardware and software computations used to construct its neuroimages but also those internal codings by which a living brain expresses itself. We interpret such images only with multiple qualifications. A recent article by Cheng and colleagues explains basic differences among some existing methods of analysis.[2]

- *Seed-based analysis* is driven by a working hypothesis. It assumes that two regions ("seeds") are believed to be in physiological communication. This approach is therefore biased by these prior assumptions and lacks an independent perspective.

- *Independent component analysis* (ICA) assumes only that the human brain has components that are independent. ICA-based methods characterize only certain differences—those correlations that are collected during a particular time series and those that represent a network-wide time series. They do not apply solely to the correlations between signals that arise from specified regions. Obviously, different modules in the brain do interact physiologically in a much more unified, coordinated manner.

- *Parcellation-based whole-brain analysis* specifies those regions (nodes) and edges (connections) that are part of a larger macroscopic brain network.

- *Voxel-level whole-brain analysis* (as illustrated in the article by Cheng et al.) is a novel, unbiased way to assess, across the whole brain, which pairs of voxels have different functional connectivity. It is used to compare the data from two large groups of subjects with data from an equally large set of controls.

Appendix F: The Enso on This Cover

> The original Buddha-nature of all living beings is like the bright moon in the sky. It appears only when the floating clouds disappear.
>
> Ch'an Master Fenyang Shanzhao (947–1024)

The *Enso*

The circle, drawn in ink with a brush is called *enso* in Japan. It has taken on a variety of interpretations.[1]

The ox-herding series of images started in eleventh-century China. It presented ten successive stages on the long spiritual path. The eighth of these showed neither the former ox nor his herder, only an empty circle. By the time this series reached Japan, the idea of an empty circle had become associated with the fully enlightened mind, empty of the fires of greed, hatred, and ignorance, cool as moonlight (chapter 23). [ZB: 443–445, 577]

The earliest known *enso* in Japan is from the brush of the master who was then the abbot of Daitokuji, Yoso Soi (1376–1458). Next to this simple circle, he wrote his telling commentary. It describes what enlightenment reveals of reality when the old veil of Self drops off: "Obvious, obvious" (*mei-reki reki*). [ZBR: 440–444]

The *Enso* on This Cover

Previous books in this series all had illustrations that depicted either a partial *enso*, the moon, or the planet Venus in the predawn sky. The original version of the *enso* on this cover is from the brush of Tanchu Terayama.[2] It was one of his 15 semicursive calligraphic works and six ink-brush paintings.

They all folded neatly into a 3 1/2 by 11 1/2 inch booklet of ink traces (*Bokuseki*). He presented it to me as I was about to leave Japan in 1989. He then explained what this traditional farewell gift symbolizes. This remarkable gift from a Zen teacher of calligraphy commemorates the previous efforts of his departing students and serves *them* as an ongoing source of inspiration.

Appendix G: Word Problems

Whenever ideas fail, men invent words.

Martin Fischer (1879–1962)

Zen has no words. When you have *satori*, you have everything.

Ch'an Master Dahui Zonggao (1089–1163)

Many lines of evidence point toward the latent dynamic powers of silence. [ZB: 367–370; 633–636] They converge to suggest that our entanglements with language in the left hemisphere hinder our awakening into *kensho-satori*. [ZBH: 30, 148–150, 155–156]

Distinctive of Zen is its distaste for wordy discourse. This is typified in the old phrase describing Zen as "the sect which does not establish words" (*Furyu monji no shu*).[1] Kobori-Roshi also warned me to beware of words that have multiple meanings. Let us consider some examples of wordy complications.

Object

The simple word *object* is very useful as a noun. [SI: 69–70] Still, there's an occasion when its use could be misleading. Let's begin over a century ago, when the pioneering neurologist Hughlings Jackson addressed the problem of consciousness. He identified (and capitalized) the two separate basic categories of our conventional "reality." The category he placed first was termed *Subject Consciousness*. He used *Subject* Consciousness to refer to every situation during which we invested *our* own Self-awareness and feelings throughout every one of our personal extensions. [ZB: 380] Jackson's use of this term *subject* was straightforward. It helped to define our dominant perspective as the only possible one that our own subjective, egocentric Self could view through its very own lens. No superficial intellectual logic probes deeply enough to grasp how extensive are the roots of this covert, *subjective* aspect of our *I-Me-Mine*. Usually, the huge bulk of this intrusive egocentric construct lies hidden from our view. Only after *kensho-satori* first cuts off its roots can the true extent of our former Subject Consciousness be appreciated. Our subjectivity becomes clear in retrospect (chapter 21). [ZBR: 261–262, 364–367]

Jackson used the contrasting term *Object Consciousness*, for the other category of reality. This second phrase was applicable to an anonymous awareness. It registered and perceived (and could re-create) the full presence of *other* things *out there*, in the external environment. Jackson used a *brick* as his example of such an external *object*. An ordinary brick (one to which we seem relatively unattached) helps us to understand what is meant by the different, dispassionate perspective that remains *objective*, allocentric, and other-centered. In this instance, objective is an adjective. It refers to the nature of reality, apart from our own feelings or thoughts.

Back in March 1982, I had not heard or read anything about Jackson's Subject and Object Consciousness. Yet a similar useful distinction leaped instantly to mind when a residual i-me-mine groped for words to define the stunning, selfless perspective it had just witnessed in *kensho*. Emerging spontaneously at that later moment was the phrase "objective vision." It signified that no "subject" (meaning no personal *I-Me-Mine*) was the intrusive agency behind that enhanced perspective of reality. No subjective lenses of my Self seemed to have refracted *that* earlier anonymous scene.

However, a semantic problem could arise. Suppose we were next to use the word *object* to describe an *item* such as an apple. We might feel much more possessive about an apple than a mere brick. Suppose this same apple-object were now subjected to the active focusing on by a *hungry* subject-person. Such a person's Self-centered, top-down, *dorsal* attention system would then be under the grasping domination of our standard form of *Subject* Consciousness. At this point, the object has become entangled in (and attached to) the limbic-tinged concepts of that hungry-feeling subject person. Why is the word *item* a more useful word than *object* as a focus of attention? Because *item* is a sufficiently neutral word choice to be perceived by either form of consciousness.

Sometimes, your goal is spoken of as your "objective." Of course, if you really intend to achieve "your" own objective, you must feel (subjectively) motivated to do so.

BuzzWords

Today, *neuroplasticity* is ensconced as a more sophisticated way to describe what we used to call learning. The word *plasticity* can sound like what happens when something is molding a brain from the *outside*. This might lead us to overlook some subtle incremental effects of long-term retraining. These help us adapt by using our brain's own intrinsic stress response systems. For example, when norepinephrine is released, it stimulates the paraventricular nucleus to release corticotrophin releasing factor. This excitatory polypeptide, in turn, reshapes adaptive responses throughout other medial regions of the cerebrum and brainstem. [ZBR: 113–120]

Lately, some research articles are using the word *conjunction* to imply that things are joined together in a unified association. Will it ever replace the old word *integration*? Integration still refers to the way separate things combine to form a complete whole.

Content/Context

In chapter 6 and figure 6.2 we discovered that these two words were used to refer to two separate categories of neural functions. However, separate words in English don't necessarily separate all the subtle boundaries and hybrid pathways that the brain uses when it codes its physiological functions.

For example, *content* begins simply enough. It refers "to specific things that are contained." But this begs the question: *in what kind of larger container are they being enclosed?*

Similarly, *context* refers to the environment. That's OK so far. But dictionaries continue this definition: "Context refers to the environment *in which something exists.*" These words point us back toward that "something" in the center—the *content* within this context that helps to define it. Soon we're in a yin/yang situation, trying to define one side unilaterally in the absence of the other.

Transcendental

In these pages, the use of a capital S in Self has a personal, localized Self-centered psychophysiological basis. This usage differs from the way the capital S tends to be used for the vast, universal "absolute witness" Self in some South Asian and transcendental meditation (TM) literature.[2] [ZBR: 184] Moreover, during TM meditation and Advita meditation [ZB: 322], the descriptions of most experiences often called "transcendental" do not appear consistent with those major, alternate, selfless states of *kensho-satori* as they have long been described in the Zen tradition (chapter 2). Instead, transcendental then usually points toward less advanced states of absorption. At these times respiration slows substantially, and the frontal EEG leads show heightened degrees of alpha 1 (8–10 cps) coherence. These episodes of heightened awareness are often described using the term, "pure consciousness." [ZBR: 237–238, 391–393]

Some long-term TM meditators notice that their enhanced sense of inner wakefulness persists even when they are asleep. These experienced trainees also have more intense rapid eye movements while dreaming. [ZBH: 84–89] Their slow-wave sleep may also show higher levels of the alpha 1 EEG activity referred to above. [SI: 296, n 11] In order to help fully characterize *kensho-satori*, a comprehensive longitudinal profile of neuroimaging changes, psychometric measurements, and self-reports will be required in addition to changes detectable in the EEG. The review by Nash and Newberg[3] helps to resolve other word problems that have plagued meditation research in the past.

Notes

Epigraph

1. I. Schloegl. *The Zen Way* (London: Sheldon Press, 1977), 16.
2. R. Bach. *Jonathan Livingston Seagull* (New York: Macmillan, 1970), 83. The Zen Buddhist Path also has ethical guidelines (*shila*).

Preface

1. J. Austin. *Zen and the Brain. Toward an Understanding of Meditation and Consciousness* (Cambridge, MA: MIT Press, 1998), 278–281, 284–286, 307.
2. J. Austin. Six Points to Ponder. *Journal of Consciousness Studies* 1999; 6(2–3): 213–216.
3. J. Austin. *Selfless Insight. Zen and the Meditative Transformations of Consciousness* (Cambridge, MA: MIT Press, 2009), 28–29, 155.
4. J. Austin. *Meditating Selflessly: Practical Neural Zen* (Cambridge, MA: MIT Press, 2011), 13–14, 43, 94, 99, 155, 157, 171, 224.
5. J. Austin. *Zen-Brain Horizons. Toward a Living Zen* (Cambridge, MA: MIT Press, 2014), xx, 77, 79, 84, 89–96, 160, 191.
6. In neurology, the word *unconscious* has a pathological connotation. It tends to imply major brain dysfunction. As used herein, *subconscious* refers to the many *normal* psychophysiological processes going on just below, or far below, our threshold of consciousness. Sometimes, like *sub*marines, these rise up and surface. When they do, we finally become aware that they exist. Chapters 10, 11, and 25 remind us that they have different levels of origin and degrees of access.

Chapter 1 Can Meditation Enhance Creative Problem-Solving Skills? A Progress Report

1. C. Prebish. *American Buddhism.* (North Scituate, MA: Duxbury Press, 1979), 38.
2. M. Baas, B. Nevicka, and F. Ten Velden. Specific Mindfulness Skills Differentially Predict Creative Performance. *Personality and Social Psychology Bulletin* 2014; 40(9): 1092–1106.
3. J. Austin. *Chase, Chance and Creativity. The Lucky Art of Novelty* (Cambridge, MA: MIT Press, 2003), 136–143, 173–185, 185–189. Does meditation enhance problem-solving skills during a period of morning *reverie*? This has yet to be studied. [CCC: 162, 184, 188]

4. These links in four books are summarized in a bracket too long for the text: [ZB: 63, 258, 621; ZBR: 103–105, 335–336, 357–358; SI: 125–130, 153–173; ZBH: 150–161, 166–167, 185].

5. F. Huang, J. Fan, and J. Luo. The Neural Basis of Novelty and Appropriateness in Processing of Creative Chunk Decomposition. *Neuroimage* 2015; 113: 122–132.

6. L. Wu, G. Knoblich, and J. Luo. The Role of Chunk Tightness and Chunk Familiarity in Problem Solving: Evidence from ERPs and fMRI. *Human Brain Mapping* 2013; 34(5): 1173–1186. Previous imaging work on insight is reviewed on page 1184 of this article.

7. R. Beaty, M. Benedek, R. Wilkins, et al. Creativity and the Default Network: A Functional Connectivity Analysis of the Creative Brain at Rest. *Neuropsychologia* 2014; 64C: 92–98. Here, the inferior parietal lobule regions appear chiefly to represent those of the angular gyrus (chapter 15).

8. X. Wan, H. Nakatani, K. Ueno, et al. The Neural Basis of Intuitive Best Next-Move Generation in Board Game Experts. *Neuroscience* 2011; 331(6015): 341–346.

9. X. Wan, D. Takano, T. Asamizuya, et al. Developing Intuition: Neural Correlates of Cognitive-Skill Learning in Caudate Nucleus. *Journal of Neuroscience* 2012; 32(48): 17492–17501.

10. X. Wan, K. Cheng, and K. Tanaka. Neural Encoding of Opposing Strategy Values in Anterior and Posterior Cingulate Cortex. *Nature Neuroscience* 2015; 18(5): 752–759.

11. R. Jung, B. Mead, J. Carrasco, et al. The Structure of Creative Cognition in the Human Brain. *Frontiers in Human Neuroscience* 2013; doi: 10.3389/fnhum.2013.00330.

12. H. Takeuchi, H. Tomita, T. Taki, et al. The Associations among the Dopamine D2 Receptor Taq1, Emotional Intelligence, Creative Potential Measured by Divergent Thinking, and Motivational State and These Associations' Sex Differences. *Frontiers in Psychology* 2015; 6: 912. doi: 10.3389/fpsyg.2015.00912.

13. M. Baas et al. Specific Mindfulness Skills, ibid. 2014. Keen powers of observation are crucial in attentive processing. However, the demand characteristics inherent in self-reports invite caution in interpreting the effects of observation in experiment III.

14. J. Austin. *Chase, Chance and Creativity.* Op. cit. 2003, 105–106, 115.

15. L. Colzato, A. Ozturk, and B. Hommel. Meditate to Create: The Impact of Focused-Attention and Open-Monitoring Training on Convergent and Divergent Thinking. *Frontiers in Psychology* 2012;

3: 116. This title nicely defines both the focused/open, and the convergent/divergent issues in creative processing.

16. M. Baas et al. *Specific Mindfulness Skills*. Op. cit. 2014.

17. D. Lippelt, B. Hommel, and L. Colzato. Focused Attention, Open Monitoring and Loving Kindness Meditation: Effects on Attention, Conflict Monitoring, and Creativity—A Review. *Frontiers in Psychology* 2014; 5: 1083.

18. J. Austin. Zen and the Brain: Mutually Illuminating Topics. *Frontiers in Psychology* 2013; 4: 784, 7–8. doi: 10.3389/fpsyg.2013.00784.

19. X. Ding, Y. Tang, R. Tang, et al. Improving Creativity Performance by Short-Term Meditation. *Behavioral and Brain Functions* 2014; 10: 9. doi: 10.1186/1744-9081-10-9.

20. X. Ding, Y. Tang, C. Cao, et al. Short-Term Meditation Modulates Brain Activity of Insight Evoked with Solution Cue. *Social Cognitive and Affective Neuroscience* 2015; 10(1): 43–49. Different remote association tests (RAT) were used before and after 10 days to avoid the influence of familiarity.

21. W. Shen, Y. Yuan, C. Liu, et al. In Search of the 'Aha!' Experience: Elucidating The Emotionality of Insight Problem-Solving. *British Journal of Psychology* 2015; doi: 10.111/bjop.12142.

22. C. Salvi, E. Bricolo, S. Franconeri, et al. Sudden Insight Is Associated with Shutting Out Visual Inputs. *Psychonomic Bulletin & Review* 2015; doi: 10.3758/s13423-015-0845-0.

23. C. Zedelius and J. Schooler. Mind Wandering "Ahas" Versus Mindful Reasoning: Alternative Routes to Creative Solutions. *Frontiers in Psychology* 2015; 6: 834.

24. E. Jauk, A. Neubauer, B. Dunst, et al. Gray Matter Correlates of Creativity Potential: A Latent Variable Voxel-Based Morphometry Study. *NeuroImage* 2015; 111: 312–320; A. Green, M. Cohen, H. Raab, et al. Frontopolar Activity and Connectivity Support Dynamic Conscious Augmentation of Creative State. *Human Brain Mapping* 2015; 36(3): 923–934; N. Mayseless and S. Shamay-Tsoory. Enhancing Verbal Creativity: Modulating Creativity by Altering the Balance Between Right and Left Inferior Frontal Gyrus with tDCS. *Neuroscience* 2015; 291: 167–176; H. Takeuchi, H. Tomite, Y. Taki, et al. Cognitive and Neural Correlates of the 5-Repeat Allele of the Dopamine D4 Receptor Gene in a Population Lacking the 7-Repeat Allele. *NeuroImage* 2015; 110C: 124–135; A. Dietrich and H. Haider. Human Creativity, Evolutionary Algorithms, and Predictive Representations: The Mechanics of Thought Trials. *Psychonomic Bulletin and Review* 2014. doi: 10.3758/s13423-014-0743-x.

25. M. Saggar, A. Zanesco, B. King, et al. Mean-field Thalamocortical Modeling of Longitudinal EEG Acquired during Intensive Meditation Training. *NeuroImage* 2015; 114: 88–104; E. Rosenberg, A. Zanesco, B. King, et al. Intensive Meditation Training Influences Emotion Reponses to Suffering. *Emotion* 2015; http://dx.doi.org/10.1037/emo000080; A. Hauswald, T. Ubelacker, S. Leske, et al. What It Means to Be Zen: Marked Modulations of Local and Interareal Synchronization during Open Monitoring Meditation. *NeuroImage* 2015; 108: 265–273; P. Faber, D. Lehmann, L. Gionotti, et al. Zazen Meditation and No-Task Resting EEG Compared with LORETA Intracortical Source Localization. *Cognitive Processing* 2015; 16(1): 87–96; P. Hemmer, Y. Guo, Y. Wang, et al. Network-Based Characterization of Brain Functional Connectivity in Zen Practitioners. *Frontiers in Psychology* 2015; 6: 603. doi: 10.3389/fpsyg.2015.00603; R. Simon and M. Engstrom. The Default Mode Network as a Biomarker for Monitoring the Therapeutic Effects of Meditation. *Frontiers in Psychology* 2015; 6: 776. doi: 10.3389/fpsyg.2015.00776; Q. Conklin, B. King, A. Zanesco, et al. Telomere Lengthening after Three Weeks of an Intensive Insight Meditation Retreat. *Psychoneuroendocrinology* 2015; doi: 10.1016/j.psyneuen.2015.07.462; F. Kurth, N. Cherbuin, and E. Luders. Reduced Age-Related Degeneration of the Hippocampal Subiculum in Long-Term Meditators. *Psychiatry Research* 2015; 232(3): 214–218. This study found a relative preservation of the left subiculum in 50 long-term meditators; E. Luders, P. Thompson, and F. Kurth. Larger Hippocampal Dimensions in Meditation Practitioners: Differential Effects in Women and Men. *Frontiers in Psychology* 2015; doi: 10.3389/fpsg.2015.00186. This study found that radial distances in the hippocampus were significantly larger in the hippocampus of 30 long-term meditators (mean duration 20.2 years) than in matched controls. In males, the findings were more evident on the left. In females, the findings were more evident on the right. It remains for further studies to clarify whether a special Tibetan technique of *nondual* meditation, as studied with fMRI with the eyes closed (1) is attributable primarily to the central portion of the precuneus and its connections [ZBR: 204–205] and (2) is also one essential component of a practical, potentially effective ("kenshogenic," as it were) approach to daily-life practice. See Z. Josipovic. Neural Correlates of Nondual Awareness in Meditation. *Annals of the New York Academy of Science* 2014; 1307: 9–18. doi: 10.1111/nyas.12261. Useful definitions for meditation, and a practical tax-

onomy for related states are now available. See the review by J. Nash and A. Newberg. Toward a Unifying Taxonomy and Definition for Meditation. *Frontiers in Psychology* 2013; doi: 10.3389/fpsyg.2013.00806.

26. R. Davidson and A. Kaszniak. Conceptual and Methodological Issues in Research on Mindfulness and Meditation. *American Psychologist* 2015; 70(7): 581–592. See also: A. Lutz, A. Jha, J. Dunne, et al. Investigating the Phenomenological Matrix of Mindfulness-Related Practices from a Neurocognitive Perspective. *American Psychologist* 2015; 70(7): 632–658; A. Harrington and J. Dunne. When Mindfulness Is Therapy: Ethical Qualms, Historical Perspectives. *American Psychologist* 2015; 70(7): 621–631; S. Dimidjian and Z. Segal. Prospects for a Clinical Science of Mindfulness-Based Intervention. *American Psychologist* 2015; 70(7): 593–620.

27. 270 Open Science Collaborators. Psychology. Estimating the Reproducibility of Psychological Science. *Science* 2015; 349(6251): aac4716. doi: 10.1126/science.aac4716.

Chapter 2. In Zen, What Does It Mean "To Be Enlightened"?

1. J. Cleary (Trans.) and T. Cleary (Ed.). *Zen Letters. Teachings of Yuanwu* (Boston, MA: Shambhala, 1994), 63. Yuanwu's comments on an earlier *koan* collection were incorporated into the popular text known as the *Blue Cliff Record*. His heir, Dahui, was so alarmed by the success of this volume that he tried to destroy copies of it.

2. K. Yamada. *Zen: The Authentic Gate* (Somerville, MA: Wisdom, 2015), 96–118. Yamada-Roshi concludes that beyond the fifth stage, the Ox-Herding Pictures are "so advanced, so removed from our actual practice" that they are of little use for most Zen students.

3. As discussed by multiple teachers throughout the entire issue entitled Enlightenment, *Inquiring Mind* 27:(1), Fall 2010, 5–25. Jack Kornfield's short introduction (pp. 5–7) is instructive. First, because he titles it "Enlightenments." Second, because he specifies the different ways that skillful teachers spontaneously express enlightenment in their *living behavior*: as hugging love; as being just where you are with a beginner's mind; as a quiet equanimity; as an energized ongoing fullness; as walking mindfully; as compassionate blessings; as sage wisdom, etc.

4. This approach assigns a much lesser weight to three *shallower* alternate states of consciousness. [ZB: 302] These brief, lesser experiences include the ordinary epiphanies, the quickenings, and the

absorptions (see reference note 5 below for the absorptions). Although these lesser states may seem impressive to the subject who experiences them, only the later deep penetrations of insight-wisdom have the potential to create a major lasting change in the person's psyche. Some neuroscientists tend to view these advanced states as confirmations of what is now being called "neuroplasticity."

5. T. Cleary. *Classics of Buddhism and Zen*, Vol. 3 (Boston, MA: Shambhala, 2001), 162, 269–314. Hakuin recognized that meditators could cultivate the absorptions by concentrating their powers of attention. He viewed the first level of absorption as the "universal mirror cognition." It could tap into "storage consciousness." This meant that such an experience could draw on memories of prior sensory events (such as a Japanese maple leaf) that had occurred earlier and been consolidated (chapter 16). It also meant that some of the items that were later retrieved from the person's subconscious could emerge into full consciousness (such as in the form of a maple leaf that was a visual hallucination). [ZB: 470–471] In addition, during an internal absorption, the total (obsidian-like) impression of space infusing the experience could seem like actually "entering a crystal world." Although some might misname this state an "illumination," the inside of such a world can also be perceived as "pitch black." Notwithstanding such word paradoxes, such a state of internal absorption might also serve as "a gate of inspiration," as it did for the present author. However, absorptions do *not* relieve a person's deep psychic afflictions.

6. Hakuin once had a dream when he was 33. It led up to an impressive closure. [ZB: 324] His mother presented him with a purple robe that had a mirror in each sleeve. The mirror in the left sleeve gave off an intense luster. After this, as he looked at everything, "it was as though I were seeing my own face." See S. Addiss, S. Lombardo, and J. Roitman (Eds.). *Zen Source Book. Traditional Documents from China, Korea, and Japan* (Indianapolis, IN: Hackett, 2008), 249. In fact, such an experience—of coidentity by Self-recognition—was only the second of *four* provisional categories that emerged when the several first-person descriptions of "oneness" were recently analyzed. [ZBR: 338–357] Large, left-sided lesions around the temporo-parietal-occipital region can be accompanied by exaggerated degrees of subjective Self-referencing. [ZBR: 15–19] These observations suggest that certain

right-hemispheric Self-referential functions could have been released as a result of the process of disinhibition (appendix B).

7. N. Waddell. *The Essential Teachings of Zen Master Hakuin* (Boston, MA: Shambhala, 1984), xviii, 33.

8. In the *Vimamsaka Sutra* (text III, 4), the person who penetrates the dharma by the direct experience of insight-wisdom deeply realizes the authenticity of the Path leading on to final release. This initial breakthrough corresponds with the earliest stage that Theravada Buddhism calls "stream entry." Again, however, this step is clearly only the first of the multiple steps on the Way toward full realization. The *Nikayas* indicate that such a "stream-enterer" could have let go of *only* the first three of the 10 fetters that still hinder full enlightenment. These first three fetters are (1) Self-centeredness; (2) doubts about the efficacy of the eightfold Path; and (3) the erroneous view that mere rituals and ascetic practices will lead to awakening. See B. Bodhi (Ed.). *In the Buddha's Words. An Anthology of Discourses from the Pali Canon* (Boston, MA: Wisdom, 2005), 86–87, 374–384.

9. T. Cleary, ibid. 298–314.

10. S. Taylor (Ed.). *Not I, Not Other Than I—The Life and Teachings of Russel Williams* (Alresford, Hants, UK, John Hunt Publisher; O Books, 2015). Chapter 8 describes Williams' state of awakening to oneness in his late twenties. Chapters 10 and 12 describe the way he experienced himself in the world afterward. "I felt a sense of oneness toward everything" (p. 94). "I only saw wholeness, with no separation, this natural state of being which was nurturing and kindness" (p. 114). If the book's title seems puzzling, consider that it describes the paradoxes of an experience that—simultaneously—combines the total absence of the old psychic Self with the recent stunning realization of coidentity with the world.

Chapter 3. Developing Traits of Character on the Way to Altruism

1. R. Aitken. *The Mind of Clover. Essays in Zen Buddhist Ethics* (San Francisco: North Point Press, 1984), 155–159. This chapter is entitled *The Way and its Virtue*.

2. D. Brooks. *The Road to Character* (New York: Random House, 2015). Eight of Brooks' major characters are from the West, mostly representing the twentieth century. The ninth is Saint Augustine (354–430). The epigraph that opens part III is from Brooks, p. 194. We return to Augustine again in chapter 25.

3. D. Brooks. ibid., 13. Eisenhower's explosive temper was infamous. Yet he also had that famous 1000-watt smile!

4. Y. Kashiwahara and K. Sonoda (Eds.). *Shapers of Japanese Buddhism* (Tokyo: Kosei, 1994), 191.

5. R. Aitken. *The Mind of Clover*, 159.

6. R. Aitken. *The Mind of Clover*, 158.

7. R. Aitken. *The Mind of Clover*, 159. Skeptical readers may doubt that long-term meditative training can sponsor the subconscious mechanisms of *implicit learning* that could accomplish any such ideal transformations. A rationale for such a gradual approach is summarized elsewhere. [MS: 169–177] It includes specific forms of "do" and "don't" suggestions. [MS: 179–189]

8. R. Aitken. *Miniatures of a Zen Master* (Berkeley, CA: Counterpoint, 2008), 26.

9. D. Suzuki. *Studies in the Lankavatara Sutra* (London: Routledge and Kegan Paul, 1930), 297.

10. M. Ricard. *Altruism. The Power of Compassion to Change Yourself and the World.* (New York: Little, Brown, 2015). This large readable volume discusses a crucial topic (text, 849 pages; notes, 135 pages).

11. A. Marsh, S. Stoycos, K. Brethel-Haurwitz, et al. Neural and Cognitive Characteristics of Extraordinary Altruists. *Proceedings of the National Academy of Sciences U.S.A.* 2014; 111(42): 15036–15041.

12. R. Boyle. *Realizing Awakened Consciousness. Interview with Buddhist Teachers and a New Perspective on the Mind* (New York: Columbia University Press, 2015), 191–204; 211–218. The teachers had been letting go of their prior *maladaptive* overconditioning, not of all prior adaptive conditioning.

13. Each Zen teacher in my own early experience had emphasized, and had exemplified in his or her own ethical behavior, the fundamental importance of *sila*—the disciplined restraint and renunciation of their maladaptive passions.

Chapter 4. The Self: A Primer

1. J. Austin. How Does Meditation Train Attention? *Insight Journal* 2009 (summer); 23: 16–22, available at bcbsdharma.org/wp-content/uploads/2013/09/09summerfullissue.pdf.

2. F. Callard and D. Margulies. What We Talk about When We Talk about the Default Mode Network. *Frontiers in Human Neurosciences* 2014; 8: 619.

3. H. Lou, M. Joensson, and K. Biermann-Ruben. Recurrent Activity in Higher Order, Modality Non-Specific Brain Regions: A Granger

Causality Analysis of Autobiographic Memory Retrieval. *PLoS One* 2011; 6(7): e22286; H. Lou, B. Luber, and J. Keenan. Self-Enhancement Processing in the Default Network: A Single-Pulse TMS Study. *Experimental Brain Research* 2012; 223(2): 177–187; H. Lou, M. Joensson, and K. Thomsen. Making Sense: Dopamine Activates Conscious Self-Monitoring through Medial Prefrontal Cortex. *Human Brain Mapping* 2015; 36(5): 1866–1877.

4. J. Austin. The Meditative Approach to Awaken Selfless Insight-Wisdom, in S. Schmidt and H. Walach (Eds.), *Meditation — Neuroscientific Approaches and Philosophical Implications* (Berlin: Springer, 2014), 23–55.

5. A similar reduction of medial prefrontal PET activity during extended deep meditation can be observed in the color PET scan frontispiece of ZB, dating from research in Japan during 1988. [ZBH: 132–133] A right hemispheric preponderance of activity is also apparent.

6. K. Garrison, T. Zeffiro, D. Scheinost, et al. Meditation Leads to Reduced Default Mode Network Activity Beyond an Active Task. *Cognitive Affective and Behavioral Neuroscience* 2015; 358, doi: 10.3758/s13415-015-0358-3.

7. Y. Dor-Ziderman, A. Berkovich-Ohana, and J. Glicksohn. Mindfulness-Induced Selflessness: A MEG Neurophenomenological Study. *Frontiers in Human Neuroscience* 2013; 7: 582, doi: 10.3389/fnhum.2013.00582.

Chapter 5. Emerging Concepts in Self–Other Relationships

1. D. Kravitz. The Ventral Visual Pathway: an Expanding Neural Framework for the Processing of Object Quality. *Trends in Cognitive Science* 2013; 17(1): 26–49.

2. J. Austin. Zen and the Brain: the Construction and Dissolution of the Self. *The Eastern Buddhist. New Series.* 1991; 24(2): 69–97.

3. J. Austin. The Meditative Approach to Awaken Selfless Insight-Wisdom, in S. Schmidt and H. Walach (Eds.), *Meditation — Neuroscientific Approaches and Philosophical Implications* (Berlin: Springer, 2014), 42–43, 53.

4. C. Gross, C. Rocha-Miranda, and D. Bender. Visual Properties of Neurons in Inferotemporal Cortex of the Macaque. *Journal of Neurophysiology* 1972; 35: 96–111. Most sampled neurons were in area TE of the monkeys' inferior temporal gyrus. After the corpus callosum and anterior commissure were sectioned, their visual response was confined to the half-field opposite that of the recording

electrode. These monkeys were studied during partial anesthesia with 70% nitrous oxide and 30% oxygen. The influence that this N_2O (and the fact that the monkeys were rigidly confined after the excitement of capture) could have had on expanding the size, shape, *and focus* of the visual fields was not commented on in this pioneering study. [ZB: 407–413; ZBR: 286, 288–290]

5. R. Tamura, T. Ono, and M. Fukuda. Recognition of Egocentric and Allocentric Visual and Auditory Space by Neurons in the Hippocampus of Monkeys. *Neuroscience Letters* 1990; 109(3): 293–298.

6. E. Rolls, N. Aggelopoulos, and F. Zheng. The Receptive Fields of Inferior Temporal Cortex Neurons in Natural Scenes. *Journal of Neuroscience* 2003; 23(1): 339–348.

7. A. Sereno, M. Sereno, and S. Lehky. Recovering Stimulus Locations Using Populations of Eye-Position Modulated Neurons in Dorsal and Ventral Visual Streams of Non-Human Primates. *Frontiers in Computational Neuroscience* 2014; 8: 28.

8. E. Rolls and T. Webb. Finding and Recognizing Objects in Nature Scenes. Complementary Computations in the Dorsal and Ventral Visual Systems. *Frontiers in Computational Neuroscience* 2014; 8: 85. The authors' computational model is biologically plausible. It shows that even though the receptive field is large, it is feasible to arrive at such an independent view in the anterior inferior cortex (area TE). How? By the converging of information at the foveal fixation point that had arrived from the monkeys' early visual pathways.

9. S. Vaziri, E. Carlson, Z. Wang, et al. A Channel for 3D Environmental Shape in Anterior Inferotemporal Cortex. *Neuron* 2014; 84(1): 55–62. The authors called the enclosed items "objects."

10. Z. Chen. Object-Based Attention. A Tutorial Review. *Attention Perception & Psychophysics* 2012; 74(5): 784–802.

11. R. Foley, R. Whitwell, and M. Goodale. The Two-Visual-Systems Hypothesis and the Perspectival Features of Visual Experience. *Consciousness and Cognition* 2015; 35: 225–233.

12. O. Baumann and J. Mattingley. Medial Parietal Cortex Encodes Perceived Heading Direction in Humans. *Journal of Neuroscience* 2010; 30(39): 12897–12901. The changes in activity were measured by the suppression that occurs during repetition. The left cerebral lateralization of the findings may reflect the fact that substantial tasks were involved in training.

13. B. Denny, H. Kober, T. Wager, et al. A Meta-Analysis of Functional Neuroimaging of Self- and Other Judgments Reveals a Spatial

Gradient for Mentalizing in Medial Prefrontal Cortex. *Journal of Cognitive Neuroscience* 2012; 24(8): 1742–1752.

14. K. Khindsa, V. Drobinin, J. King, et al. Examining the Role of the Temporo-Parietal Network in Memory, Imagery, and Viewpoint Transformations. *Frontiers in Human Neuroscience* 2014; 8: 709. Compare with reference 27.

15. C. Parkinson, S. Liu, and T. Wheatley. A Common Cortical Metric for Spatial, Temporal, and Social Distance. *Journal of Neuroscience* 2014; 34(5): 1979–1987. The region studied would represent more the *angular gyrus* (BA 39) than the supramarginal gyrus (BA 40).

16. Y. Mah, M. Husain, G. Rees, et al. Human Brain Lesion-Deficit Inference Remapped. *Brain* 2014; 137 (Pt 9): 2522–2531.

17. A. Hillis, M. Newhart, and J. Heidler. Anatomy of Spatial Attention: Insights from Perfusion Imaging and Hemispatial Neglect in Acute Stroke. *Journal of Neuroscience* 2005; 25(12): 3161–3167. The actual coexisting subcortical infarcts did not correlate with any type of neglect. One patient's allocentric neglect was mentioned in association with underperfusion of the right posterior inferior temporal gyrus (BA 37). Another patient's allocentric neglect was associated with right angular gyrus underperfusion. Line bisection tests did not reliably correlate with ego- or allo- neglect. One patient had both forms of neglect.

18. D. Mort, P. Malhotra, and S. Mannan. The Anatomy of Visual Neglect. *Brain* 2003; 126: 1986–1997. All posterior cerebral artery patients with neglect also had left homonymous field defects. Six other patients with lesions within the right posterior cerebral artery distribution had no neglect. I thank Professor Hans-Otto Karnath for calling this article to my attention when we met in 2014.

19. C. Bird, P. Malhotra, A. Parton, et al. Visual Neglect after Right Posterior Cerebral Artery Infarction. *Journal of Neurology, Neurosurgery and Psychiatry* 2006; 77(9): 1008–1012.

20. C. Grimsen, H. Hildebrant, and M. Fahle. Dissociation of Egocentric and Allocentric Coding of Space in a Series of Visual Search Tasks after Only Right Middle Cerebral Artery Stroke. *Neuropsychologia* 2008; 46(3): 902–914.

21. Y. Yue, W. Song, S. Huo, et al. Study on the Occurrence and Neural Bases of Hemispatial Neglect with Different Reference Frames. *Archives of Physical Medicine and Rehabilitation* 2012; 93(1): 156–162. Major cortical infarctions occurred in these cases, not underperfusion. The Ota gap detection tests were used. In no instance was

a right gap detected incorrectly. All patients with egocentric ne-glect detected the left gap incorrectly.

22. G. Humphreys, C. Gillebert, M. Chechlacz, et al. Reference Frames in Visual Selection. *Annals of the New York Academy of Science* 2013; 1296: 75–87. This review discusses the complex nature of the is-sues in this field.

23. C. Goeke, S. Kornpetpanee, M. Koster, et al. Cultural Background Shapes Spatial Reference Frame Proclivity. *Scientific Reports* 2015; 5: 11426.

24. A. Torok, T. Nguyen, O. Kolozsvari, et al. Reference Frames in Vir-tual Spatial Navigation Are Viewpoint Dependent. *Frontiers in Human Neuroscience* 2014; 8: 686.

25. A. Saj, Y. Cojan, B. Musel, et al. Functional Neuro-Anatomy of Egocentric versus Allocentric Space Representation. *Neurophysiol-ogie Clinique* 2014; 44(1): 33–40. This egocentric task also activated a part of the parahippocampus.

26. C. Lin, T. Chiu, and K. Gramann. EEG Correlates of Spatial Orien-tation in the Human Retrosplenial Complex. *Neuroimage* 2015; 120: 123–132.

27. M. Boccia, F. Nemmi, and C. Guariglia. Neuropsychology of Environmental Navigation in Humans: Review and Meta-Analysis of fMRI Studies in Healthy Participants. *Neuropsychology Review* 2014; 24(2): 236–251. Compare with reference 14.

28. A. Ekstrom, A. Arnold, and G. Iaria. A Critical Review of the Al-locentric Spatial Representation and its Neural Underpinnings: Toward a Network-Based Perspective. *Frontiers in Human Neuro-science* 2014; 8: 803. doi: 10.3389/fnhum.2014.00803. The last quota-tions come from page 12.

29. D. Li, H. Karnath, and C. Rorden. Egocentric Representations of Space Co-Exist with Allocentric Representations: Evidence from Spatial Neglect. *Cortex* 2014; 58: 161–169.

Chapter 6. Early Distinctions between Self and Other, Focal and Global, Are Coded in the Medial Temporal Lobe

1. J. Knierim, J. Neunuebel, and S. Deshmukh. Functional Correlates of the Lateral and Medial Entorhinal Cortex: Objects, Path Integra-tion and Local-Global Reference Frames. *Philosophical Transactions of the Royal Society of London. Series B: Biological Sciences* 2013; 369: 20130369.

2. T. Hartley, C. Lever, N. Burgess, J. O'Keefe. Space in the Brain: How the Hippocampal Formation Supports Spatial Cognition.

Philosophical Transactions of the Royal Society of London. Series B: Biological Sciences 2013; 369(1635): 20120510.

3. T. Ono, K. Nakamura, M. Fukuda, et al. Place Recognition Responses of Neurons in Monkey Hippocampus. *Neuroscience Letters* 1991; 121(1–2): 194–198.

4. E. Moser, Y. Roudi, M. Witter, et al. Grid Cells and Cortical Representation. *Nature Reviews Neuroscience* 2014; 15(7): 466–481.

5. J. Taube. The Head Direction Signal. Origins and Sensory-Motor Integration. *Annual Review of Neuroscience* 2007; 30: 181–207. The term *head direction cells* sounds Self-centered from the beginning. What's to prevent readers from concluding that these two words don't actually point back to the spatial coordinates of their very own Self-centered head? Less ambiguous terms, such as *angular* direction cells or *goal* directional cells, could help anchor a concept *of direction* per se. See M. Drexel, A. Preidt, E. Kirchmair, et al. Parvalbumin Interneurons and Calretinin Fibers Arising from the Thalamic Nucleus Reuniens Degenerate in the Subiculum after Kainic Acid-Induced Seizures. *Neuroscience* 2011; 189: 316–329.

6. R. Robertson, E. Rolls, and P. Georges-Francois. Head Direction Cells in the Primate Pre-Subiculum. *Hippocampus* 1999; 9(3): 206–219.

7. T. Bjerknes, R. Langston, I. Kruge, et al. Coherence among Head Direction Cells before Eye Opening in Rat Pups. *Current Biology* 2015; 25(1): 103–108.

8. F. Sargolini, M. Fyhn, T. Hafting, et al. Conjunctive Representation of Position, Direction, and Velocity in Entorhinal Cortex. *Science* 2006; 312(5774): 758–762.

9. F. Carpenter, D. Manson, K. Jeffery, et al. Grid Cells Form a Global Representation of Connected Environments. *Current Biology* 2015; 25(9): 1176–1182.

10. Figure 6.2 leaves out the circuitry involving the dentate gyrus, the CA3 cells, and those bidirectional pathways that link the reuniens nucleus of the thalamus both with the CA1 neurons of the ventral hippocampus and those of the medial prefrontal cortex (chapter 8).

11. B. Kraus, M. Brandon, R. Robinson, et al. During Running in Place, Grid Cells Integrate Elapsed Time and Distance Run. *Neuron* 2015; 88(3): 578–589.

12. A. Maaso, D. Berron, L. Libby, et al. Functional Subregions of the Human Entorhinal Cortex. *Elife* 2015; doi: 10.7554/eLife.06426.

13. T. Navarro Schrodder, K. Haak, N. Zaragoza Jimenez, et al. Functional Topography of the Human Entorhinal Cortex. *Elife* 2015; doi: 10.7554/eLife.06738.

14. M. Ritchey, L. Libby, and C. Ranganath. Cortico-Hippocampal System Involved in Memory and Cognition: The PMAT Framework. *Progressive Brain Research* 2015; 219: 45–64.

15. J. Aggleton. Multiple Anatomical Systems Embedded within the Primate Medial Temporal Lobe: Implications for Hippocampal Function. *Neuroscience and Biobehavioral Reviews* 2012; 36(7): 1579–1596.

16. M. Chadwick and H. Spiers. A Local Anchor for the Brain's Compass. *Nature Neuroscience* 2014; 17(11): 1436–1437.

17. M. Chadwick, A. Jolly, D. Amos, et al. A Goal Directional Signal in the Human Entorhinal/Subicular Region. *Current Biology* 2015; 25(1): 87–92.

Chapter 7. Remindfulness

1. R. W. Emerson. *Essays by Ralph Waldo Emerson* (New York: Perennial Library, Harper & Row, 1926), 99. This statement occurs in a long chapter titled "Spiritual Laws" (pp 93–120). It was published in 1841, in Emerson's first series of essays. In Emerson's perspective at that time, we humans were mere "parts or particles" of the enveloping Universal Whole. The true calling of each person's character would be shaped by how well he or she were listening, at this lowly level, to the wordless guidance flowing spontaneously from this infinite, overarching Universe. Ten years earlier, in 1831, the first major railroad tracks had pushed north, on the Mohawk and Hudson line. One legend from the early railroading years was that the Native Americans could tell when the next "iron horse" was coming by bending over, applying their lower ear to the track, and listening for the faint rumble of the train in the distance. Whether this legend entered into his perspective is not known.

2. The Mindfulness Movement; What Does It Mean for Buddhism? An interview by J. Wilks with M. Blacker, B. Boyce, D. Winston, and T. Goodman. *Buddhadharma*. Spring 2015, 46–55. See also reference 26 in chapter 1 for a critique of the current research in mindfulness.

3. Dalai Lama. *Beyond Religion: Ethics for a Whole World* (Boston, MA: Houghton Mifflin Harcourt, 2011), 109.

4. The memory expert, Endel Tulving, was said to have estimated that there could be 256 different types of memory (!) Each had its own laws of encoding, consolidation and retrieval. The remindfulness under discussion in part III serves as a general term useful in pointing toward one category of these memory functions. The two examples in chapter 10 illustrate how these memory functions were experienced in a long-term Zen meditative context. During retrieval, every recollection function doesn't necessarily have to depend solely on how much conscious and subconscious attention the person had "paid" to the earlier event. In fact, some mechanisms could be accessed at random. These could operate more or less independently of how casually or intently the attentive processing had been conducted while the original incident was being encoded. [ZB: 390–391]

5. D. Levinson, E. Stoll, S. Kindy, et al. A Mind You Can Count On: Validating Breath Counting as a Behavioral Measure of Mindfulness. *Frontiers in Psychology* 2014; 5: 1202. doi: 10.3389/fpsyg. The authors noted that mindfulness also includes the awareness that some word-thoughts, even though unrelated to the task, can coincide during the same present moment. Working memory capacity was tested by "holding seven letters in memory while doing math," as in the automated operation span test (OSPAN).

6. D. King, M. de Chastelaine, R. Elward, et al. Recollection-Related Increases in Functional Connectivity Predict Individual Differences in Memory Accuracy. *Neuroscience* 2015; 35(4): 1763–1772. Recollection is defined here as "The processes that support retrieval of qualitative information about a prior event." The recollection skills of 88 subjects in their 20s were monitored by fMRI during three different behavioral paradigms based on associations to words, pictures, and coins.

7. H. Lu, Y. Song, and M. Xu. The Brain Structure Correlates of Individual Differences in Trait Mindfulness: A Voxel-Based Morphometry Study. *Neuroscience* 2014, doi: 10.1016/j.neuroscience. 2014.04.051.

8. E. Luders, P. Thompson, and F. Kurth. Larger Hippocampal Dimensions in Meditation Practitioners: Differential Effects in Men and Women. *Frontiers in Psychology* 2015; 6:186. doi: 10.3389/fpsyg.2015.00186. This report is the latest in a long series of important related studies from this group. [ZBH: 167, 169]

9. E. Luders, N. Cherbuin, and F. Kurth. Forever Young(er): Potential Age-Defying Effects of Long-Term Meditation on Gray Matter

Atrophy. *Frontiers in Psychology* 2015; 5: 1551. doi: 10.3389/fpsyg.2014.01551. The authors emphasize that this is a cross-sectional, not longitudinal, study. They present a conservative multifactorial discussion of their findings. The cutoff at the young age of 77 qualifies how much comfort a nonagenarian might draw from the data. The report by Gefen et al. is more encouraging. In their 31 living "Super Agers" over 80, the right rostral anterior cingulate region was thicker than that in controls. Moreover, their five autopsied cases (ages 81–95) all showed a greater density of von Economo neurons in the anterior cingulate region. See T. Gefen, M. Peterson, S. Papastefan, et al. Morphometric and Histologic Substrates of Cingulate Integrity in Elders with Exceptional Memory Capacity. *Journal of Neuroscience* 2015; 35(4): 1781–1791. [SI: 138–139, 239; MS: 102, 215 n. 14]

10. In Theravada Buddhism, the Pali term *vipassana* is translated as insight. Vipassana meditative practices in the West have come to be identified as "insight meditation." The usage of this term does not exclude the fact that insights can develop at any point along the Path of those who practice Mahayana Buddhism. For example, open-eyed Zen meditative practices also cultivate insights into one's true nature, into the nature of all things as they really are, and intuitions in general.

11. W. Hasenkamp, C. Wilson-Mendenhall, E. Duncan, et al. Mind Wandering and Attention During Focused Meditation: A Fine-Grained Temporal Analysis of Fluctuating Cognitive States. *NeuroImage* 2012; 59(1): 750–760.

12. A. Dijksterhuis. First Neural Evidence for the Unconscious Thought Process. *Social Cognitive and Affective Neuroscience* 2013; 8: 845–846. Thoughts are definable as intangible phenomena that tend to carry subjective weight. In general, intuitive processes solve problems subconsciously, usually without our needing to hear word thoughts at that instant.

13. A. Abbott. Unconscious Thought Not So Small after All. *Nature* 2015: 517(7536): 537–538. doi: 10.1038/517537a; T. Stafford. The Perspectival Shift: How Experiments on Unconscious Processing Don't Justify the Claims Made for Them. *Frontiers in Psychology* 2014; 5: 1067. doi: 10.3389/fpsyg.2014.01067.

14. J. Cresswell, J. Bursley, and A. Satpute. Neural Reactivation Links Unconscious Thought to Decision-Making Performance. *Social Cognitive and Affective Neuroscience* 2013; 8: 863–869.

15. M. Abadie, L. Waroquier, and P. Terrier. Gist Memory in the Unconscious-Thought Effect. *Psychological Science* 2013; 24(7): 1253–1259.

16. A. Vlassova, C. Donkin, and J. Pearson. Unconscious Information Changes Decision Accuracy but Not Confidence. *Proceedings of the National Academy of Sciences USA* 2014; 111(45): 16214–16218. The lack of confidence is in favor of a form of subconscious processing.

17. J. Li, Y. Zhu, Y. Yang. The Merits of Unconscious Thought in Rule Detection. *Public Library of Science One* 2014; 9(8): e106557. doi: 10.1371/journal.pone.0106557.

18. J. Zhou, C. Zhou, J. Li, et al. Cognitive Style Modulates Conscious but Not Unconscious Thought: Comparing the Deliberation-Without-Attention Effect in Analytics and Wholists. *Consciousness and Cognition* 2015; 36: 54–60.

19. G. Vallee-Tourangeau, M. Abadie, and F. Vallee-Tourangeau. Interactivity Fosters Bayesian Reasoning Without Instruction. *Journal of Experimental Psychology-General* 2015; 144(3): 581–603.

20. C. Stillman, E. Gordon, J. Simon, et al. Caudate Resting Connectivity Predicts Implicit Probabilistic Sequence Learning. *Brain Connectivity* 2013; 3(6): 601–610.

21. A. Schaefer, D. Margulies, G. Lohmann, et al. Dynamic Network Participation of Functional Connectivity Hubs Assessed by Resting-State fMRI. *Frontiers in Human Neuroscience* 2014; 8: 195.

22. X. Chai, N. Ofen, J. Gabrieli, et al. Development of Deactivation of the Default-Mode Network During Episodic Memory Formation. *NeuroImage* 2014; 84: 932–938.

23. X. Chai, N. Ofen, J. Gabrieli, et al. Selective Development of Anticorrelated Networks in the Intrinsic Functional Organization of the Human Brain. *Journal of Cognitive Neuroscience* 2014; 26(3): 501–513.

24. G. Galli. What Makes Deeply Encoded Items Memorable? Insights into the Levels of Processing Framework from Neuroimaging and Neuromodulation. *Frontiers in Psychiatry* 2014; 5: 61.

Chapter 8. A Remindful Route through the Nucleus Reuniens

1. R. Vertes, S. Linley, and W. Hoover. Limbic Circuitry of the Midline Thalamus. *Neuroscience and Biobehavioral Reviews* 2015; pii: S0149-7634(15)00016-0. doi: 10.1016/j.neurobiorev.2015.01.014. The route through the reuniens is more conspicuous in rodents than in primates.

2. S. Duss, T. Reber, J. Hanggi, et al. Unconscious Relational Encoding Depends on Hippocampus. *Brain* 2014; 137(Pt. 12): 3355–3370. doi: 10.1093/brain/awu270. Notably, activations occurred *within* the hippocampus when masked faces and words were *jointly* encoded subconsciously. No hippocampal activations occurred when *single* words were encoded subconsciously. Results of this kind suggest that the hippocampus plays a role in associational memories.

3. J. Aggleton. Multiple Anatomical Systems Embedded Within the Primate Medial Temporal Lobe: Implications for Hippocampal Functions. *Neuroscience & Biobehavior Reviews* 2012; 36(7): 1579–1596. Substantial connections also link the prefrontal cortex with the parahippocampus. These signaling processes appear to help confer a sense of familiarity and to be associated with the retrieval of memories.

4. G. Pergola and B. Suchan. Associative Learning Beyond the Medial Temporal Lobe: Many Actors on the Memory Stage. *Frontiers in Behavioral Neuroscience* 2013; 7: 162. doi: 10.3389/fnbeh.2013.00162. Figure 1 in this article proposes that the lateral side of the entorhinal cortex has reciprocal connections with the perirhinal cortex. The medial side has reciprocal connections with the parahippocampal cortex. In this article, arrows in figures 2 and 3 also point to the location of the reuniens nucleus.

5. Our consciousness cannot identify each deep, widespread source of our instantly coded recollections. The spoor of their constituent messages vanishes too fast as it nears the surface of even higher levels of introspection. This doesn't prevent researchers from trying to follow memory traces with MEG (appendix C).

6. M. Drexel, W. Hoover, and R. Vertes. Collateral Projections from Nucleus Reuniens of Thalamus to Hippocampus and Medial Prefrontal Cortex in the Rat: A Single and Double Retrograde Fluorescent Labeling Study. *Brain Structure and Function* 2011; 217(2): 191–209.

7. J. Prasad, A. Abela, and Y. Chudasama. Midline Thalamic Reuniens Lesions Improve Executive Behaviors. *Neuroscience* 2016; doi: 10.1016/j.neuroscience.2016.01.071.

8. F. Davoodi, F. Motamedi, E. Akbari, et al. Effect of Reversible Inactivation of Reuniens Nucleus on Memory Processing in Passive Avoidance Task. *Behavioral Brain Research* 2011; 221(1): 1–6.

9. A. Pereira de Vasconcelos and J. Cassel. The Nonspecific Thalamus: A Place in a Wedding Bed for Making Memories Last. *Neuroscience & Biobehavioral Reviews* 2015; 54: 175–196.

10. M. Zust, P. Colella, and T. Reber. Hippocampus Is Place of Interaction between Unconscious and Conscious Memories. *Public Library of Science One* 2015; 10(3): e0122459. doi: 10.1371/journal. pone.0122459.

11. R. Polania, W. Paulus, and M. Nitsche. Noninvasively Decoding the Contents of Visual Working Memory in the Human Prefrontal Cortex Within High-Gamma Oscillatory Patterns. *Journal of Cognitive Neuroscience* 2012; 23(2): 304–314.

12. R. Vertes. Major Diencephalic Inputs to the Hippocampus: Supramammillary Nucleus and Nucleus Reuniens. Circuitry and Function. *Progress in Brain Research* 2015; 219: 121–144. The reuniens is believed to amplify signals from CA 3 and the entorhinal cortex to CA 1. The supramammillary nucleus is believed to amplify signals from the entorhinal cortex to the dentate gyrus. It is part of the ascending system from the brainstem that contributes to the hippocampal theta rhythm.

Chapter 9. A Disorder Called Transient Global Amnesia

1. T. Bartsch, J. Dohring, and A. Rohr. CA 1 Neurons in the Human Hippocampus Are Critical for Autobiographical Memory, Mental Time Travel and Autonoetic Consciousness. *Proceedings of the National Academy of Sciences of the United States of America* 2011; 108(42): 17562–17567. The fact that one designated neurologist was always "on call for this study 24/7" helps to explain the excellence of this patient-based research. In this TGA study all acute retrograde deficits of episodic memory involved only the patients' most recent time periods. The *hyperintense* DWI lesions were considered to "mirror an impaired cellular metabolism." Chapter 17 suggests that several neurochemical mechanisms of NO• are consistent with such an impression. The lesions were detected in 2 mm slices using both diffusion-weighted imaging and T2-weighted images. They were confirmed by *hypointense* lesions in the apparent diffusion coefficient (ADC) maps. These were maximal at three to five days. The lateral distribution of these hippocampal lesions is noteworthy. It is a potential clue to the nature of the emerging functions of those lateral and medial paths that we began to follow in chapter 6 as they led forward and converged into the hippocampal formation. The patients in this particular

report were considered to have suffered a disproportionate loss of autobiographical memories related to their personal selves. This contrasted with the lesser loss of those memories related to their semantic knowledge of facts.

2. A. Ropper and R. Brown. *Adams and Victor's Principles of Neurology*, 8th ed. (New York: McGraw-Hill, 2005), 379–380. TGA can recur. Several series suggest that the patients as a group are more likely to report a past history of migraine and hypertension. Discussed elsewhere in the following pages of *Zen Brain Reflections* are the relationships among migraine (306–312), nitric oxide (279–288), and pertinent neuropsychological symptoms in various states of consciousness (449–450, 455). As a first-year neurology resident with Raymond Adams, Miller Fisher, and Maurice Victor in 1949–1950, I recall vividly the excitement prompted by the admissions of the early TGA patients.

3. J. Arena and A. Rabinstein. Transient Global Amnesia. *Mayo Clinic Proceedings* 2015; 90(2): 264–272. Future research must define the precise mechanisms that converge to briefly disconnect the hippocampus and other regions in transient global amnesia.

4. T. Bartsch, K. Alfke, R. Stingele, et al. Selective Affection of Hippocampal CA-1 Neurons in Patients with Transient Global Amnesia without Long-Term Sequelae. *Brain* 2006; 129(Pt 11): 2874–2884.

5. K. Han, A. Chao, F. Chang, et al. Obstruction of Venous Drainage Linked to Transient Global Amnesia. *Public Library of Science One* 2015; 10(7): e0132893. doi: 10.1371/journal.pone.0132893.

6. Y. Kang, E. Kim, J. Kim, et al. Time of Flight MR Angiography Assessment Casts Doubt on the Association between Transient Global Amnesia and Intracranial Jugular Venous Reflux. *European Radiology* 2015; 25(3): 703–709.

7. M. Gupta, M. Kantor, C. Tung, et al. Transient Global Amnesia with a Unilateral Infarction of the Fornix: Case Report and Review of the Literature. *Frontiers in Neurology* 2015; 5: 291. doi: 10.3389/fneur.2014.00291.

8. N. Giannantoni, G. Lacidogna, A. Broccolini, et al. Thalamic Amnesia Mimicking Transient Global Amnesia. *Neurologist* 2015; 19(6): 149–152.

9. To clarify the mechanisms of *kensho's* initial and "afterglow" phases, future research teams could benefit from combining the assets of the serial DWI, T2, fMRI connectivity, and EEG/MEG approaches. The recommendations for doing this would proceed on the same urgent basis as does the current protocol of comprehensive

emergency care for the neurological patient who is suffering from an acute stroke-in-evolution.

10. A. Forster, M. Griebe, and A. Gass. Diffusion-Weighted Imaging for the Differential Diagnosis of Disorders Affecting the Hippocampus. *Cerebrovascular Diseases* 2012; 33(2): 104–115.

11. T. Bartsch et al. CA 1 Neurons, 2011. In this instance, the disproportionate loss of this kind of personal Self-relational information was referred to using the term *autonoetic*.

12. Plausible leads to the pathophysiology of this singular memory disorder could now be followed during a preclinical research protocol that includes (1) examining the intimate blood supply to, and the vascular drainage from, this lateral CA 1 region with specific regard to a delayed phase during which (a) neurochemical and (b) congestion-vascular mechanisms could increase its local vascular permeability; and (2) correlating this information with the local potential for elevated toxic levels of NO• to occur during an earlier phase as the result of the glutamate → NO• → cyclic GMP cascade. [ZBR: 279–288, 306–312] The advent of 7-Tesla imaging in human subjects offers an important avenue to resolve the relevant preclinical and clinical issues (appendix D).

13. M. Peer, N. Nitzan, I. Goldberg, et al. Reversible Functional Connectivity Disturbances during Transient Global Amnesia. *Annals of Neurology* 2014; 75(5): 634–643. The authors considered that (the earliest use of) diffusion-weighted imaging might detect unilateral lesions in only about 20 percent of such acute patients, and bilateral lesions in only about 12 percent of the acute patients. However, on clinical grounds, a TGA patient's acute loss of memory is so severe that one might anticipate that this patient's *physiological abnormality* would indeed be *bilateral*, not unilateral. Clearly, in addition to basic preclinical research, more serial, longitudinal, fMRI data, DWI data, and MEG data would be helpful to clarify these clinical paradoxes.

14. T. Bartsch and S. Arzy. Human Memory: Insights into Hippocampal Networks in Epilepsy. *Brain* 2014; 137: 1856–1857. The authors note that attempts to minimize the role of the CA 1 field as an explanation for the declarative memory deficits of patients with chronic epilepsy must be qualified. Why? Because these chronic patients have had many years during which their brains can develop compensatory reorganizations.

15. T. Bartsch, J. Dohring, S. Reuter, et al. Selective Neuronal Vulnerability of Human Hippocampal CA1 Neurons: Lesion Evolution,

Temporal Course, and Pattern of Hippocampal Damage in Diffusion-Weighted MR Imaging. *Journal of Cerebral Blood Flow and
Metabolism* 2015; doi. 10.1038/jcbfm.2015.137. As a potential mechanism for cytotoxic damage and vasogenic edema, nitric oxide
could be added to this list of pathogenic mechanisms.

Chapter 10. Remindful Zen: An Auditory "Altar Ego"?

1. O. Singleton, B. Holzel, M. Vangel, et al. Change in Brainstem
    Gray Matter Concentration Following a Mindfulness-Based Inter-
    vention Is Correlated with Improvement in Psychological Well-
    Being. *Frontiers in Human Neuroscience* 2014; 8: 33. doi: 10.3389/
    fnhum.2014.00033. Even an eight-week MBSR course can increase
    the gray matter concentration in two symmetrical clusters on each
    side of the pons. These regions can enter into particular mood,
    arousal, *and* sensory mechanisms. In this dorsal pons reside the
    locus coeruleus (a major source for norepinephrine), the raphe
    pontis (a source for serotonin), and the sensory nucleus of the tri-
    geminal nerve (the major source of sensory information about our
    head). [ZB: 201–204, 205–208; ZBR: 477–478]
    Moreover, the tensor tympani muscle (attached to the malleus of
    the middle ear) receives both its sensory and motor innervation
    from the sensory nucleus and adjacent motor nucleus of this fifth
    cranial nerve. [ZB: 454] When a nitric oxide donor molecule pre-
    cipitates migraine headaches, it increases the PET scan activations
    in this dorsal portion of the pons. [ZBR: 306–312]

2. W. Britton, J. Lindahl, B. Cahn, et al. Awakening is Not a Meta-
    phor: The Effects of Buddhist Meditation Practices on Basic Wake-
    fulness. *Annals of New York Academy of Science* 2014; 1307: 64–81.

3. I was looking forward to singing this song *slowly*, two weeks
    hence, in barbershop harmony with other family members at a
    birthday celebration. Moreover, the actual and metaphoric "light"
    of the "silvery moon" had long been a personal source of multiple
    associations. [ZBR: 405–407, 414–447, 459–461]

4. I had taken that Thursday afternoon off to select a sylvan bench
    site to be a surprise present for my significant other, in gratitude.
    [MS: 140–145]

5. T. Cleary. *Zen Essence. The Science of Freedom* (Boston, MA: Shamb-
    hala, 1989), cf. 76. Yongming added that "You'll get bogged down
    along the way if you can't see past the bad habits that have taken
    root within your own psyche."

6. Certain kinds of silent ongoing speech inside our head can be helpful guides that point to our different underlying emotions. This important principle needs no introduction for those viewers of the latest Disney animated film, "*Inside Out.*" We hear different human voices in this PIXAR film. They express the presence of Anger, Fear, Disgust, Sadness—and can-do Joy—in a little girl's inner speech. Her emotional word thoughts arise in "Headquarters." This "control center" is much easier for the audience to understand than most networks identified in these pages. However, labeling our own negative emotions does imply that we have developed enough "distance" to recognize that they are temporary mind states. This understanding of impermanence helps to defuse their emotional charge. [SI: 9]

7. M. Perrone-Bertolotti, L. Rapin, J. Lachaux, et al. What Is That Little Voice Inside My Head? Inner Speech Phenomenology, Its Role in Cognition Performance, and Its Relation to Self-Monitoring. *Behavioral Brain Research* 2014; 261: 220–239.

8. L. Johns, K. Kompus, M. Connell, et al. Auditory Verbal Hallucinations in Persons with and without a Need for Care. *Schizophrenia Bulletin* 2014; Suppl 4: S255–264. doi: 10.1093/schbul/sbu005. Certain phenomena point toward a benign form of auditory hallucinations. They include a neutral or pleasant content, a low intensity of associated distress, infrequency, short duration, and low impact on the person's overall level of functioning.

9. B. Alderson-Day, S. McCarthy-Jones, and C. Fernyhough. Hearing Voices in the Resting Brain: A Review of Intrinsic Functional Connectivity Research on Auditory Verbal Hallucinations. *Neuroscience & Biobehavioral Reviews* 2015; 55: 78–87.

10. K. Hugdahl. Auditory Hallucinations: A Review of the ERC "VOICE" Project. *World Journal of Psychiatry* 2015; 5(2): 193–209.

11. D. Sivarao. The 40-Hz Auditory Steady-State Response: A Selective Biomarker for Cortical NMDA Function. *Annals of the New York Academy of Science* 2015; 1344: 27–36. This NMDA receptor (N-methyl-*d*-aspartate) is activated by the release of glutamate or aspartate (appendix B). When the brains of meditators and migraineurs are stimulated, their lower sensory thresholds need to be approached with appropriate caution.

12. D. Leitman, D. Wolf, J. Ragland, et al. "It's Not What You Say, But How You Say It": A Reciprocal Temporo-Frontal Network for Affective Prosody. *Frontiers in Human Neuroscience* 2010; 4: doi: 10.3389/fnhum.2010.00019.

13. A. At, L. Spierer, and S. Clarke. The Role of the Right Parietal Cortex in Sound Localization: A Chronometric Single Pulse Transcranial Magnetic Stimulation Study. *Neuropsychologia* 2011; 49(9): 2794–2797. Right parietal TMS blocks the subjects' spatial discrimination of sounds, especially if these sound stimuli had started 20 milliseconds earlier (more directly) out in the *left* half of space. In contrast, the spatial discrimination of auditory stimuli presented out in the right half of space is not blocked unless the sounds had been presented 80 milliseconds earlier.

14. A. Awad and H. Luders. Hypnopompic Seizures. Epileptic Disorders 2010; 12(4): 270–274. Very rarely, patients with intractable epilepsy have brief seizures that *in themselves* stimulate those arousal mechanisms that then *awaken* the patient. These seizures occur during the second stage of sleep, often with rhythmic theta activity. The present author has no clinical seizures. His daytime sleep EEG shows the usual K complexes and no seizure discharges. (Incidentally, no other family members are known to have had hallucinations, or to have been diagnosed with schizophrenia.).

15. J. McGrath, S. Saha, A. Al-Hamzawi, et al. Psychotic Experiences in the General Population: A Cross-National Analysis Based on 31,261 Respondents from 18 Countries. *Journal of the American Medical Association Psychiatry* 2015; 72(7): 697–705.

16. B. Krakvik, F. Laroi, A. Kalhovde, et al. Prevalence of Auditory Verbal Hallucinations in a General Population: A Group Comparison Study. *Scandinavian Journal of Psychology* 2015; doi: 10.1111/sjop.12236.

17. C. Choong, M. Hunter, and P. Woodruff. Auditory Hallucinations in Those Populations That Do Not Suffer from Schizophrenia. *Current Psychiatry Reports* 2007; 9(3): 206–212.

18. L. Lewis-Hanna, M. Hunter, T. Farrow, et al. Enhanced Cortical Effects of Auditory Stimulation and Auditory Attention in Healthy Individuals Prone to Auditory Hallucinations during Partial Wakefulness. *NeuroImage* 2011; 57(3): 1154–1161.

19. A preliminary informal survey suggests that the simpler bell hallucinations could be more common after the eighth decade in a few other seniors, none of whom have meditated.

20. A search for the perceptual memory underlying these early morning "sound bytes" might begin by correlating their onset with certain types and subtypes of rhythmic events that arise normally in our brains during each 24-hour cycle. For a recent review, see P. Halasz, R. Bodiz, L. Parrino, et al. Two Features of Sleep Slow

Waves: Homeostatic and Reactive Aspects—From Long Term to Instant Sleep Homeostasis. *Sleep Medicine* 2014; 15: 1184–1195; Y. Sun, J. Pita-Almenar, C. Wu, et al. Biphasic Cholinergic Synaptic Transmission Controls Action Potential Activity in Thalamic Reticular Nucleus Neurons. *Journal of Neuroscience* 2013; 33(5): 2048–2059.

Chapter 11. Following an Auditory Stimulus, Then "Seeing the Light"

1. J. Lindahl, C. Kaplan, E. Winget, et al. A Phenomenology of Meditation-Induced Light Experiences: Traditional Buddhist and Neurobiological Perspectives. *Frontiers in Psychology* 2014; 4: 973.
2. The subject disappeared from follow-up after describing this remarkable story. In an effort to preserve his anonymity, I am obligated to delete all of the usual identifying details. His past history is negative for migraine, epilepsy, or head trauma. He had tried psilocybin mushrooms once, but had no other psychedelic exposure.
3. J. Lindahl et al. Phenomenology of Meditation-Induced Light Experiences.
4. L. Kuan Yu (Charles Luk). *The Secrets of Chinese Meditation. Self-Cultivation by Mind Control as Taught in the Ch'an, Mahayana and Taoist Schools of China.* (York Beach, ME: Samuel Weiser, 1984) 15–42. In the imaginative plot of the *Surangama* Sutra, Manjusuri specifies the single most effective way to arrive at an advanced internal state of imperturbability. It consists of "disengaging the organ of hearing from its object—sound—and then directing that organ into the stream of concentration." (p. 16, 17). (These words seem reminiscent of some components in E.G.'s description.) Moreover, in this Sutra, *Avalokitesvara* (*Quan-yin*) is said to favor meditation that involves one's sense of hearing (p. 33). Indeed, "the profound uncreative [inherent] power [of this deep state of realization] enables 14 kinds of fear*lessness* to be bestowed upon all living beings." [ZB: 567–570] This indicates that audition is an avenue into an essential component of *kensho-satori*. For the record, why would this imaginative sutra, said to have been translated as late as 705 (p. 228), be cited in this chapter? Simply because it testifies to the historical fact that *audition*—more than the other sensory avenues—was recognized as playing a major role in the belief system then in use by the Chinese Buddhists. The sutra is not in the Pali canon. It is not a historically accurate account of an

assembly at which the Buddha was actually present with his 25 bodhisattvas.

5. J. Rauschecker. An Expanded Role for the Dorsal Auditory Pathway in Sensorimotor Control and Integration. *Hearing Research* 2011; 271(1–2): 16–25.

6. R. Remedios, N. Logothetis, and C. Kayser. A Role of the Claustrum in Auditory Scene Analysis by Reflecting Sensory Change. *Frontiers in Systems Neuroscience* 2014; 8: 44.

7. E. Halgren, J. Sherfey, A. Irimia, et al. Sequential Temporo—Fronto—Temporal Activation during Monitoring of the Auditory Environment for Temporal Patterns. *Human Brain Mapping* 2010; 32(8): 1260–1276.

8. C. Dumcommun, C. Michel, S. Clarke, et al. Cortical Motion Deafness. *Neuron* 2004; 43(6): 765–777.

9. X. Zhang, Q. Zhang, X. Hu, et al. Neural Representations of Three-Dimensional Acoustic Space in the Human Temporal Lobe. *Frontiers in Human Neuroscience* 2015; 9: 203. doi: 10.3389/fnhum.2015.00203. Only "near space" was being studied—57 cm in front and 57 cm behind the subjects. Right-versus-left discriminations were much more accurate than the others. Multivariate pattern analysis was necessary to interpret the temporal lobe fMRI data.

10. J. Clery, O. Guipponi, C. Wardak, et al. Neuronal Bases of Peripersonal and Extrapersonal Spaces, Their Plasticity and Their Dynamics: Knowns and Unknowns. *Neuropsychologia* 2015; 70: 313–326. doi: 10.1016/j.neuropsychologia.2014.10.022.

11. Z. Yue, Y. Jiang, Y. Li, et al. Enhanced Visual Dominance in Far Space. *Experimental Brain Research* 2015; doi: 10.1007/x00221-015-4353-2.

12. N. Van der Stoep, S. Van der Stigchel, T. Nijboer, et al. Audiovisual Integration in Near and Far Space: Effects of Changes in Distance and Stimulus Effectiveness. *Experimental Brain Research* 2015; doi: 10.1007/s00221-015-4248-2.

13. K. Yamada. *The Gateless Gate. The Classic Book of Zen Koans* (Somerville, MA: Wisdom, 2015) 8.

14. Japanese Buddhism uses the phrase, *jiji muge hokkai* to refer in general to the harmonious interdependence and oneness of all life.

Chapter 12. Turning

1. J. Cleary and T. Cleary (Trs. and Eds.) *Zen Letters. Teachings of Yuan Wu* (Boston, MA: Shambhala, 1994), VI.

2. A. Ferguson. *Zen's Chinese Heritage. The Masters and Their Teachings* (Boston, MA: Wisdom, 2000), 77–85. Baizang's illustrous teacher Mazu Daoyi lived from 709 to 788. Mazu had also called out to a departing person. Then, when that person turned, he too had asked, "*What* is it?" (pp. 70–71).

3. Cleary and Cleary. Zen Letters, 25.

4. L. Ungerleider and M. Mishkin. Two Cortical Visual Systems, in *Analysis of Visual Behavior,* ed. D. Ingle et al. (Cambridge, MA: MIT Press, 1982), 549–586.

5. Yamada-Roshi's writings show that he was familiar with the words "turning toward" when they are used to point toward such concepts as the first of five successively deeper levels of awakening. In this context the phrase implies that the Zen trainee's *every* behavior turns literally (from bending at the waist, for example) *and* figuratively in those directions that will express the truth of the Dharma. See K. Yamada. *Zen. The Authentic Gate* (Somerville, MA: Wisdom, 2015), 101–104.

6. K. Yamada. *The Gateless Gate. The Classic Book of Zen Koans* (Somerville, MA: Wisdom, 2015), case 7, 42–43.

7. Near-infrared spectroscopy (NIRS) offers a method to test how standing up (and other behaviors impossible in a fMRI scanner) could influence our perceptions of Self. Perhaps NIRS could yield some cortical explanations for why I have to stop and ask myself, "*What* did I intend to accomplish when I moved into the next room?"

8. J. Houston. The Psychenaut Program. An Exploration Into Some Human Potentials. *Journal of Creative Behavior* 1973; 7: 253–278. Our vestibular system is coded to respond to different movements. The semicircular canals react to turning, rotational movements. The utricle and saccule react to movements against or toward gravity. This information reaches the pons via the vestibular branch of the eighth cranial nerve. It is relayed to our vestibular cortex and posterior insula.

9. A. Braverman. *Mud and Water: A Collection of Talks by the Zen Master Bassui* (San Francisco, CA: North Point Press, 1989), 45.

10. K. Gramann, H. Muller, E. Eick, et al. Evidence of Separable Spatial Representations in a Virtual Navigation Task. *The Journal of Experimental Psychology: Human Perception and Performance* 2005; 31(6): 1199–1223.

11. T. Baker. C. Holroyd. The Topographical N170: Electrophysiological Evidence of a Neural Mechanism for Human Spatial Navigation. *Biological Psychology* 2013; 94(1): 90–105.

12. T. Baker, A. Umemoto, A. Krawitz, et al. Rightward-Biased Hemodynamic Response of the Parahippocampal System During Virtual Navigation. *Scientific Reports* 2015; 5: 9063. doi: 10.1038/srep09063.

13. It remains for more precise testing to determine how such a context-sensitive "highlighting" could apply to those theta and "inside" data functions that tend to be more associated with the medial entorhinal cortex (MEC) and/or to the seemingly more "landmark," "outside" data functions of the LEC (chapter 6).

14. The greater fMRI signal in the right parahippocampal region is in keeping with other evidence that our right hemisphere is the more "adroit" in its spatial orientation. [SI: 31–32, 72] The right posterior insula is also more responsive to turning signals from the semicircular canals. [ZBR: 97] When the task for 13 adults (10 right, 3 left handed) was to start from the center of a large room and then run as fast as possible around its periphery, most subjects showed the trend (a 0.708 probability) for turning *counterclockwise* ($p <$ 0.01). See Y. Toussaint and J. Fagard. A Counterclockwise Bias in Running. *Neuroscience Letters* 2008; 442(1): 59–62. Therefore, to the degree that this left-turning bias normally exists, then the process of overcoming it by turning to the right will require a different qualitative and quantitative kind of response.

15. At many ice skating rinks, you can directly experience your own habitual preference to skate counterclockwise. You'll also discover that this easier-to-turn-*counter*clockwise bias is perpetuated culturally at many racetracks in those countries where humans, horses, and dogs run longer distances. Asian countries offer exceptions to such generalizations. Moreover, it is an ancient Indian and Tibetan Buddhist custom to walk in a clockwise direction around sacred sites. This ritual had no obvious neural basis. Instead, according to cultural *tradition*, it is said to relate back to the apparent course of the sun—rising over in the East, then moving to the South, before setting in the West.

Chapter 13. Revisiting *Kensho*, March 1982

1. C. Jung. *Psychology and Religion. West and East*, Vol. 11, Bollingen Series 20 (New York: Pantheon, 1958), 546.

2. First, it became necessary to assemble the host of plausible mechanisms that (by converging on different levels of the thalamus) could accomplish the major shifts required for this earlier working hypothesis. The author's condensed discussion of these steps began in Your Self, Your Brain, and Zen. *Cerebrum. The Dana Forum on Brain Science* 2003; 5(1): 47–66; and in Selfless Insight-Wisdom. A Thalamic Gateway, in *Measuring the Immeasurable. The Scientific Case for Spirituality*, D. Goleman (Ed.) (Boulder, CO: Sounds True, 2008), 211–230. Next, a neural model illustrating the physiology of the triggering shift needed to be developed. [SI: 114, figure 7]

3. Why is the clarity of this immediate retrograde memory of more than passing interest? Because it contrasts with that opposite tendency, occurring in transient global amnesia, for a short, retrograde phase of amnesia to be added to the other acute ongoing phenomena. A recent behavioral study of 119 normal subjects suggests that one's memory for seemingly inconsequential earlier details can be enhanced *retroactively* after conceptually related associations are activated emotionally. See J. Dunsmoor, V. Murty, L. Davachi, et al. Emotional Learning Selectively and Retroactively Strengthens Memories for Related Events. *Nature* 2015; 520(7547): 345–348.

4. N. Bolognini, S. Convento, M. Fusaro, et al. The Sound-Induced Phosphene Illusion. *Experimental Brain Research* 2013; 231(4): 469–478. Sounds boost visual cortical responses.

5. T. Baker, A. Umemoto, A. Krawitz, et al. Rightward-Biased Hemodynamic Response of the Parahippocampal System during Virtual Navigation. *Scientific Reports* 2015; 5: 9063. doi: 10.1038/srep09063.

6. J. Viviano and K. Schneider. Interhemispheric Interactions of the Human Thalamic Reticular Nucleus. *Journal of Neuroscience* 2015; 35(5): 2026–2032.

7. Z. Kafkas and D. Montaldi. Two Separate, but Interacting, Neural Systems for Familiarity and Novelty Detection: A Dual-Route Mechanism. *Hippocampus* 2014; 24(5): 516–527. "Familiarity" becomes a semantic paradox the instant that a complete absence of Self occurs.

8. J. Schomaker and M. Meeter. Short- and Long-Lasting Consequences of Novelty, Deviance and Surprise on Brain and Cognition. *Neuroscience Biobehavioral Reviews* 2015; 55: 268–279. doi: 10.1016/j.neurobiorev.2015.05.002.

9. J. Cleary and T. Cleary (Trs. and Eds.) *Zen Letters. Teachings of Yuan Wu* (Boston, MA: Shambhala, 1994), 54.

Chapter 14. A Mondo in Clinical Neurology

1. See M. Chechlacz and G. Humphreys. The Enigma of Balint's Syndrome: Neural Substrates and Cognitive Deficits. *Frontiers of Human Neuroscience* 2014; 8: 123. doi: 10.3389/fnhum.2014.00123. One obvious problem centers on the notion that more dorsal lesions would interfere mostly with focal vision, *not* global vision. However, syndrome implies that more than one mechanism is present.

2. B. Bodhi. (Ed.) *In The Buddha's Words. An Anthology of Discourses from the Pali Canon.* (Boston, MA: Wisdom, 2005), 214–215. (*Udana* 6:4; 67–69) The blind man who feels only the tusk argues that it's just like a plowshare. The man who feels an ear contends that it's just like a big basket for separating the chaff from the grain, etcetera.

3. C. Thomas, K. Kveraga, E. Huberle, et al. Enabling Global Processing in Simultanagnosia by Psychophysical Biasing of Visual Pathways. *Brain* 2012; 135(Pt 5): 1578–1585.

4. R. Ptak. The Frontoparietal Attention Network of the Human Brain: Action, Saliency, and a Priority Map of the Environment. *Neuroscientist* 2012; 18(5): 502–515.

5. A. Khan, M. Prost-Lefebvre, R. Salemme, et al. The Attentional Fields of Visual Search in Simultanagnosia and Healthy Individuals: How Object and Space Attention Interact. *Cerebral Cortex* 2015; Apr 2. pii: bhv059. The patient (I.G.) had extensive bilateral posterior damage to the superior parietal lobule and also to the angular gyrus.

6. A. Sripati and C. Olson. Representing the Forest Before the Trees: A Global Advantage Effect in Monkey Inferotemporal Cortex. *Journal of Neuroscience* 2009; 29(24): 7788–7798. These nerve cells were located in the ventral bank of the left superior temporal sulcus (STS), (chapter 15), or in the left inferior temporal gyrus. Of the 123 neurons tested in 2 monkeys, 17 percent were selective for a *global* configuration of the test image; 24 percent were sensitive to the identity of the *local* elements in a hierarchical display. Notice that if both types of nerve cells were to participate in a fully integrated co-activation, the result would seem capable of conferring higher-level interpretive properties clearly and directly *throughout all features of an entire scene.* (chapter 5) These dual attributes would seem ideal for representing, instantly, the kinds of high-level allocentric features—infused with a sense of meaning—that contribute to the "big picture" in *kensho-satori.* [ZB: 249]

That they *do* serve this purpose is a hypothesis not easily testable at this time.

7. K. Dalrymple, J. Barton, and A. Kingstone. A World Unglued: Simultanagnosia as a Spatial Restriction of Attention. *Frontiers of Human Neuroscience* 2013; 7: 145.

8. C. Mevorach, L. Shalev, R. Green, et al. Hierarchical Processing in Balint's Syndrome: A Failure of Flexible Top-Down Attention. *Frontiers of Human Neuroscience* 2014; 8:113, doi: 10.3389/fnhum.2014.00113.

9. D. Balslev, B. Odoj, J. Rennig, et al. Abnormal Center-Periphery Gradient in Spatial Attention in Simultanagnosia. *Journal of Cognitive Neuroscience* 2014; 26(12): 2778–2788.

10. R. Ptak and J. Fellrath. Exploring the World with Balint Syndrome: Biased Bottom-Up Guidance of Gaze by Local Saliency Differences. *Experimental Brain Research* 2014; 232(4): 1233–1240. This patient had multiple bilateral lesions and oculomotor apraxia.

11. S. Lomber, S. Malhotra, and J. Sprague. Restoration of Acoustic Orienting into a Cortically Deaf Hemifield by Reversible Deactivation of the Contralesional Superior Colliculus: The Acoustic "Sprague Effect." *Journal of Neurophysiology* 2007; 97(2): 979–993.

12. K. Schneider. Subcortical Mechanisms of Feature-Based Attention. *Journal of Neuroscience* 2011; 31(23): 8643–8653.

13. B. Corneil and D. Munoz. Overt Reponses During Covert Orienting. *Neuron* 2014; 82(6): 1230–1243.

14. R. Krauzlis, L. Lovejoy, and A. Zenon. Superior Colliculus and Visual Spatial Attention. *Annual Review of Neuroscience* 2013; 36: 165–182.

15. S. Fisher and J. Reynolds. The Intralaminar Thalamus—An Expressway Linking Visual Stimuli to Circuits Determining Agency and Action Selection. *Frontiers in Behavioral Neuroscience* 2014; 8: 115.

16. A. Celeghin, B. de Gelder, and M. Tamietto. From Affective Blindsight to Emotional Consciousness. *Consciousness* 2015; http://dx.doi.org/10.1016/j.concog.2015.05.007.

Chapter 15. Two Key Gyri, a Notable Sulcus, and the Wandering Cranial Nerve

1. R. Murray, M. Debbane, P. Fox, et al. Functional Connectivity Mapping of Regions Associated with Self- and Other-Processing. *Human Brain Mapping* 2015; 36(4): 1304–1324. Such research begs

the question: *who* is implicit in *our* sense of meaning and in each of *our* movements?

2. M. Seghier. The Angular Gyrus: Multiple Functions and Multiple Subdivisions. *Neuroscientist* 2013; 19(1): 43–61. The dorsal/ventral parcellation is an empirical finding. It is not clear why this observation departs from the usual dorsal-ego/ventral-allo trend.

3. R. Ferreira, S. Gobel, M. Hymers, et al. The Neural Correlates of Semantic Richness: Evidence from an fMRI Study of Word Learning. *Brain and Language* 2015; 143: 69–80.

4. A. Price, M. Bonner, J. Peelle, et al. Converging Evidence for the Neuroanatomic Basis of Combinatorial Semantics in the Angular Gyrus. *Journal of Neuroscience* 2015; 35(7): 3276–3284. Many cognitive/association functions of the angular gyrus could be jointly referable to its links with the central precuneus and with the medial and lateral prefrontal cortex. See D. Margulies, J. Vincent, C. Kelly, et al. Precuneus Shares Intrinsic Functional Architecture in Humans and Monkeys. *Proceedings of the National Academy of Sciences USA* 2009; 106(47): 20069.

5. S. Lee, J. Booth, and T. Chou. Developmental Changes in the Neural Influence of Sublexical Information on Semantic Processing. *Neuropsychologia* 2015; 73: 25–34.

6. Z. Douglas, B. Maniscalco, M. Hallett, et al. Modulating Conscious Movement Intention by Noninvasive Brain Stimulation and the Underlying Neural Mechanisms. *Journal of Neuroscience* 2015; 35(18): 7239–7255. This is a sham-controlled study.

7. B. Clemens, S. Jung, G. Mingoia, et al. Influence of Anodal Transcranial Direct Current Stimulation (tDCS) over the Right Angular Gyrus on Brain Activity During Rest. *Public Library of Science One* 2014; 9(4): e95984: doi: 10.1371/journal.pone.009584. The authors describe the right angular gyrus in their subjects as "covering BA 40." They include it in their "task positive network," not in their default mode network. Their tDCS experiments were performed without sham controls.

8. L. Cloutman, R. Binney, D. Morris, et al. Using in Vivo Probabilistic Tractography to Reveal Two Segregated Dorsal 'Language-Cognitive' Pathways in the Human Brain. *Brain and Language* 2013; 127(2): 230–240.

9. C. Ranganath and M. Ritchey. Two Cortical Systems for Memory-Guided Behavior. *Nature Reviews Neuroscience* 2012; 13(10): 713–726. The authors propose that the perirhinal cortex is more specialized for our memory of individual objects (or items),

whereas the parahippocampus supports our memory for scenes, spatial layout (or contexts). Moreover, the perirhinal cortex projects more into the *lateral* entorhinal cortex (LEC) and forward into an *anterior* temporal system. In contrast, the parahippocampus projects more into the medial entorhinal cortex (MEC) and up into a posterior medial system that includes the navigational functions of the retrosplenial cortex (chapter 6).

10. L. Wang, R. Mruczek, M. Arcaro, et al. Probabilistic Maps of Visual Topography in Human Cortex. *Cerebral Cortex* 2014; 25(10), doi: 10.1093/cercor/bhu277. Figure 12A illustrates the upper field and foveal representations.

11. K. Weiner and K. Grill-Spector. Neural Representations of Faces and Limbs Neighbor in Human High-Level Visual Cortex: Evidence for a New Organization Principle. *Psychological Research* 2013; 77(1): 74–97. By specifying the existence of this *lateral* occipital → temporal visual processing pathway, the authors expand on the previous standard version of only one dorsal ("Where?") pathway and only one ventral ("What?") pathway. In a sense, this lateral occipital → temporal pathway cortex could be seen as fulfilling a more *intermediate* ("equatorial") role in its style of multimodal processing. As information flows from its lateral occipital sulcus toward the middle occipital gyrus, it conveys codes for position, form, and shape. Next, in the posterior inferior temporal sulcus, it processes information about position and motion. Then, when it relays up into the middle temporal gyrus, it carries information about form, visual dynamics, touch, action, and language. Their figure 11 summarizes this useful model of an intermediate pathway that begins laterally. This differs from the pathway that courses through the undersurfaces of the temporal lobe (figure 15.1).

12. J. Bastin, J. Vidal, S. Bouvier, et al. Temporal Components in the Parahippocampal Place Area Revealed by Human Intracerebral Recordings. *Journal of Neuroscience* 2013; 33(24): 10123–10131. The patients were candidates for the surgical relief of their seizure disorder.

13. J. Aggelton. Multiple Anatomical Systems Embedded within the Primate Medial Temporal Lobe: Implications for Hippocampal Function. *Neuroscience Biobehavioral Reviews* 2012; 36(7): 1579–1596.

14. B. Deen, K. Koldewyn, N. Kanwisher, et al. Functional Organization of Social Perception and Cognition in the Superior Temporal Sulcus. *Cerebral Cortex* 2015; doi: 10.1093/cercor/bhv111.

15. T. Flack, T. Andrews, M. Hymers, et al. Responses in the Right Posterior Superior Temporal Sulcus Show a Feature-Based Response to Facial Expression. *Cortex* 2015; 69: 14–23.

16. S. Korb, S. Fruhholz, and D. Grandjean. Reappraising the Voices of Wrath. *Social Cognitive and Affective Neuroscience* 2015; 10(12): 1644–1660. doi. 10.1093/scan/nsv051. The left superior frontal gyrus was activated and became coupled with the right auditory cortex during the reappraisal intended to *increase* the emotional response. However, during the reappraisal intended to *decrease* the emotional response, the activated networks included the medial frontal gyrus, the posterior parietal regions (R > L) and the auditory regions bilaterally.

17. C. Pernet, P. McAleer, M. Latinus, et al. The Human Voice Areas: Spatial Organization and Inter-Individual Variability in Temporal and Extra-Temporal Cortices. *Neuroimage* 2015; doi: 10.1016/j.neuroimage.2015.06.650.

18. R. Ruffoli, F. Giorgi, C. Pizzanelli, et al. The Chemical Neuroanatomy of Vagus Nerve Stimulation. *Journal of Chemical Neuroanatomy* 2011; 42(4): 288–296.

19. K. Vonck, R. Raedt, J. Naulaerts, et al. Vagus Nerve Stimulation ... 25 Years Later! What Do We Know about the Effects on Cognition? *Neuroscience & Biobehavioral Reviews* 2014; 45: 63–71.

20. E. Frangos, J. Ellrich, and B. Komisaruk. Non-Invasive Access to the Vagus Nerve Central Projections via Electrical Stimulation of the External Ear: fMRI Evidence in Humans. *Brain Stimulation* 2015; 8(3): 624–636.

21. J. Austin and S. Takaori. Studies of Connections between Locus Coeruleus and Cerebral Cortex. *The Japanese Journal of Pharmacology* 1976; 26(2): 145–160.

Chapter 16. Paradox: The Maple Leaf Way Up in Ambient Space

1. J. Hackett. *The Zen Haiku and Other Zen Poems of J. W. Hackett* (Tokyo: Japan Publications, 1983), 13.

2. References 4 and 8 in chapter 5 suggest that stress-related changes that drive an aroused, excitable state also have the dynamic potential to blend a laser-like one-pointed focusing into a vast spatial global context.

Chapter 17. The Nitric Oxide Connection

1. R. Santos, C. Lourenco, A. Ledo, et al. Nitric Oxide Inactivation Mechanisms in the Brain: Role in Bioenergetics and Neurodegen-

eration. *International Journal of Cell Biology* 2012; article number 391914, http://dx.doi.org/10.1155/2012/391914.

2. C. Bechade, S. Colasse, M. Diana, et al. NOS2 Expression Is Restricted to Neurons in the Healthy Brain but Is Triggered in Microglia upon Inflammation. *Glia* 2014; 62(6): 956–963, doi: 10.1002/glia.22652. This report describes NO• in mouse brain.

3. C. Lourenco, R. Santos, R. Barbosa, et al. Neurovascular Coupling in Hippocampus Is Mediated via Diffusion by Neuronal-Derived Nitric Oxide. *Free Radical Biology and Medicine* 2014; 73: 421–429.

4. Z. Katusic and S. Austin. Endothelial Nitric Oxide: Protector of a Healthy Mind. *European Heart Journal* 2014; 34(14): 888–894. The release of these neuronal and endothelial-generated NO• molecules is the essential signaling event that promotes many normal biochemical reactions. A genetically determined defecit in endothelial NO• can have vascular consequences in the hippocampus and elsewhere (see X. Tan, Y. Xue, T. Ma, et al. Partial eNOS Deficiency Causes Spontaneous Thrombotic Cerebral Infarction, Amyloid, Angiopathy and Cognitive Impairment. *Molecular Neurodegeneration* 2015; 10: 24, doi: 10.1186/s13024-015-0020-0. However, the excessive release of NO• can be cytotoxic. These delayed changes are currently being associated with aspects of Alzheimer disease, Parkinson disease, and Lewy body dementia (see T. Nakamura, O. Prikhodko, E. Pirie, et al. Aberrant Protein S-Nitrosylation Contributes to the Pathophysiology of Neurodegenerative Diseases. *Neurobiology of Disease* 2015; 84: 99–108, doi: 10.1016/j.nbd.2015.03.017.)

5. C. Lourenco, N. Ferreira, R. Santos, et al. The Pattern of Glutamate-Induced Nitric Oxide Dynamics in Vivo and Its Correlation with nNOS expression in Rat Hippocampus, Cerebral Cortex and Striatum. *Brain Research* 2014; 1554: 1–11.

6. Z. Jing, X. Wei, S. Wang, et al. The Synergetic Effects of Nitric Oxide and Nicotinic Acetylcholine Receptor on Learning and Memory of Rats. *Sheng Li Xue Bao Journal* 2014; 66(3): 307–314.

7. J. Gasulla and D. Calvo. Enhancement of Tonic and Phasic GABAergic Currents Following Nitric Oxide Synthase Inhibition in Hippocampal CA 1 Pyramidal Neurons. *Neuroscience Letters* 2015; 590: 29–34.

8. M. Moosavi, L. Abbasi, A. Zarifkar, et al. The Role of Nitric Oxide in Spatial Memory Stages, Hippocampal ERK and CaMKII Phosphorylation. *Pharmacology, Biochemistry and Behavior* 2014; 122: 164–172.

9. T. Moraes-Neto, A. Scopinho, C. Biogone, et al. Involvement of Dorsal Hippocampus Glutamatergic and Nitrergic Neurotransmission in Autonomic Responses Evoked by Acute Restraint Stress in Rats. *Neuroscience* 2014; 258: 364–373. The enlarged visual fields in the early primate studies (chapter 5) are phenomena explainable in large part on the basis of similar hyperexcitability discharges.

10. A. Samdani, T. Dawson, and V. Dawson. Nitric Oxide Synthase in Models of Focal Ischemia. *Stroke* 1997; 28(6): 1283–1288.

11. D. Fujikawa. The Role of Excitotoxic Programmed Necrosis in Acute Brain Injury. *Computational and Structural Biotechnology Journal* 2015; 13: 212–221.

12. J. Austin. *Chase, Chance and Creativity. The Lucky Art of Novelty* (Cambridge, MA: MIT Press, 2003), 63, 108, 110.

13. Nothing ventured, nothing gained.

14. Y. Gunduztepe, S. Mit, and E. Gecioglu. The Impact of Acupuncture Treatment on Nitric Oxide (NO) in Migraine Patients. *Acupuncture Electrotherapy Research* 2014; 39(3–4): 275–283.

15. X. Tan, Y. Xue, X. Wang, et al. Partial eNOS Deficiency Causes Spontaneous Thrombotic Cerebral Infarction, Amyloid Angiopathy and Cognitive Impairment. *Molecular Neurogeneration* 2015; 10(1): 24. doi: 10.1186/s13024-015-0020-0.

Chapter 18. "Pop-Out"

1. T. Ossandon, J. Vidal, C. Ciumas, et al. Efficient "Pop-Out" Visual Search Elicits Sustained Broadband Gamma Activity in the Dorsal Attention Network. *Journal of Neuroscience* 2012; 32(10): 3414–3421.

2. T. Hayakawa, S. Miyauchi, N. Fujimaki, et al. Information Flow Related to Visual Search Assessed Using Magnetoencephalography. *Brain Research. Cognitive Brain Research* 2003; 15(3): 285–295. Although pop-out is easiest to study visually, it can also be demonstrated in audition.

3. T. Ossandon, et al. Efficient "Pop-Out" Visual Search, p. 2102.

4. M. Watson, A. Brennan, A. Kinstone, et al. Looking versus Seeing: Strategies Alter Eye Movements During Visual Search. *Psychonomic Bulletin and Review* 2010; 17(4): 543–549. Circles with one gap are part of the Ota test described in chapter 5. Discriminating between circles with one, two, or no gaps requires accurate, sustained visual attention. The passively instructed students spent more time gazing at the center of the display before making their first saccade. The researchers interpreted this as consistent with the

greater emphasis placed on (more passive) actual "seeing," as contrasted with (more active) searching while "looking." However, this could have shortened the average distance to the average target.

5. Y. Erez and G. Yovel. Clutter Modulates the Representation of Target Objects in the Human Occipitotemporal Cortex. *Cognitive Neuroscience* 2014; 26(3): 490–500.

6. A. Vallesi. Monitoring Mechanisms in Visual Search: An fMRI Study. *Brain Research* 2014; 1579: 65–73.

Chapter 19. Keeping Your Eye on the Ball

1. S. Green, with G. McAlpine. *The Way of Baseball. Finding Stillness at 95 mph* (New York: Simon & Schuster, 2011), 149. Green seems at home with attentional skills even though he has not established a formal Zen practice. Notice how his chapters' titles condense each of the following key Zen topics in a few well-chosen words: stillness, space and separation, awareness, ego, presence, the Zone, nonattachment, and gratitude. Green retired in 2007, at the age of 34. His career batting average was 0.282. He had hit 328 home runs and batted-in 1071 runs.

2. P. Jackson and H. Delehanty. *Sacred Hoops. Spiritual Lessons of a Hardwood Warrior* (New York: Hyperion, 1995) Raised in a Pentecostal household, Jackson practiced open-eyed Zen meditation and "learned to trust the moment" (p. 51). After playing professional basketball himself, he started as coach of the Chicago Bulls in 1989. His (Zen-oriented) teaching methods soon brought results: the Bulls went on to win three successive National Basketball Association championships (from 1991 to 1993) and a fourth in 1996.

3. G. Wulf and R. Lewthwaite. Effortless Motor Learning? An External Focus of Attention Enhances Movement Effectiveness and Efficiency, in *Effortless Attention. A New Perspective in the Cognitive Science of Attention and Action*, ed. B. Bruya (Cambridge, MA: MIT Press, 2010), 75–101.

4. A body scan is a deliberate, top-down, internal focusing on parts of one's own somatic Self. For novice meditators who lack a sense of familiarity with their own body image in action, its risk/benefit ratio starts out relatively low. For someone preoccupied with somatic, hypochondriacal complaints, a body scan would be contraindicated. Many other excellent ways exist to train top-down attention that do not reinforce one's own internal Self-referential

constructs. Inhabiting the real world of Nature outdoors is one of these methods (chapters 10 and 24). None of the author's formative Zen training in the mid-seventies involved a body scan.

5. K. Erickson, R. Prakash, M. Voss, et al. Aerobic Fitness Is Associated with Hippocampal Volume in Elderly Humans. *Hippocampus* 2009; 19: 1030–1039. These elderly subjects were not meditators.

Chapter 20. What Is Living Zen?

1. T. Cleary. *Zen Essence, The Science of Freedom* (Boston, MA: Shambhala, 1989), 57–58. Despite Dahui's criticism of the "silent illumination school" of Zen (the Caodong school in China; Soto, later in Japan), the Caodong master Hongzhi still requested that Dahui be the executor of his final affairs after he died.

2. G. Fox, J. Kaplan, H. Damasio, et al. Neural Correlates of Gratitude. *Frontiers in Psychology* 2015; 6: 1491. doi: 10.3389/fpsyg.2015.01491.

Chapter 21. Sometimes, Zen Is "For the Birds"

1. A. Ferguson. *Zen's Chinese Heritage. The Masters and Their Teachings* (Boston, MA: Wisdom, 2000), 331.

2. P. Coyote. It! It! It! *Shambhala Sun* 2015; 24(4): 64–71. Coyote was later informed that the unseen bird issuing these shrieks was thought to be a "Camp Jay." Bird watchers may wonder about this common label. The calls of the blue-colored Western Scrub Jay are harsh, loud, rising "*shreeeenks.*" The gray-black Clark's Nutcracker is larger and more fearless, but it makes longer, rising "*shraaaaaaa*" calls.

3. Parts of this Coyote account are comparable with the selfless perfection, eternity, fearlessness, and nonintervention of another *kensho* narrative. [ZB: 536–548, 542–544; ZBR: 407–410, 414–415, 428–429] However, that London episode (condensed in chapter 13) had *no* loud bird call or other major sudden sensory stimulus that could have triggered it.

4. B. Leach. *Hamada, Potter* (Tokyo: Kodansha International, 1975), 64–65, 97. Many thanks to my dharma Brother, Milton Moon, the illustrious Zen potter and author, for his letter of May 16, 2014 that pointed me toward this remarkable book.

5. There are pros and cons to word labels. It is certainly helpful to step back from a troubling state of mind, to identify it, to objectify it, and to attach a simple world-label to it (e.g., "fear," "rumination," "planning"). [SI: 9] It is also useful to know the common

names of different birds, plants, trees, mountains, etc., at least to the degree that this does help us appreciate their individual ecological role in Nature. However, those birder competitions that drive some to "win" by achieving the longest lists of bird names are another matter.

Chapter 22. Basho, the *Haiku* Poet

1. R. Blyth. *A History of Haiku, Volume I: From the Beginnings up to Issa* (Japan: Hokuseido Press, 1963), 9.
2. J. Miller, R. Hayden, R. O'Neal (Eds.) *The United States in Literature* (Glenville, IL: Scott, Foresman and Company, 1976), 211 M. This summarizes the closing lines of a poem, entitled "Poetry," by Marianne Moore (1887–1972). Her *Collected Works* (1952) received the Pulitzer Prize, the National Book Award, and the Bollinger Prize.
3. M. Ueda (Ed.). *Basho and His Interpreters. Selected Hokku with Commentary* (Stanford, CA: Stanford University Press, 1991), 9.
4. T. Oseko (Ed.). *Basho's Haiku. Literal Translations for Those Who Wish to Read the Original Japanese Text, with Grammatical Analysis and Explanatory Notes* (Tokyo: Maruzen, 1990), 43.
5. M. Ueda. *Basho and His Interpreters*, 3.
6. L. Stryk (Tr.). *On Love and Barley: Haiku of Basho* (Harmondsworth, Middlesex, England: Penguin, 1985) 9.
7. R. Blyth, *A History of Haiku*, 4, 5.
8. R. Blyth, *A History of Haiku*, 8.
9. R. Blyth, *A History of Haiku, Volume II: From Issa up to the Present* (Japan: Hokuseido Press, 1964), 332.
10. J. Austin. Zen and the Brain: Mutually Illuminating Topics. *Frontiers in Psychology* 2013; 4: 1–9, doi: 10.3389/fpsyg.2013.00789.
11. R. Blyth. *A History of Haiku, Volume II*, 349.
12. R. Aitken. *A Zen Wave. Bashō's Haiku and Zen* (Washington, DC: Shoemaker & Hoard, 2003), as described by the poet W. S. Merwin, ix. (This is a revision of the 1978 edition.).
13. R. Aitken, ibid., xviii—xix.
14. R. Aitken, ibid., 4.
15. R. Aitken, ibid., 5.
16. R. Aitken, ibid., 6.
17. R. Aitken, ibid., 5.
18. S. Addiss. *The Art of Haiku. Its History through Poems and Paintings by Japanese Masters* (Boston, MA: Shambhala Publications, 2012), 172.

19. H. Shirane. *Traces of Dreams. Landscape, Cultural Memory, and the Poetry of Basho* (Stanford, CA: Stanford University Press, 1998), 30–51.

20. D. Suzuki. *Zen and Japanese Culture. Bollingen Series LXIV* (Princeton, NJ: Princeton University Press, 1973), 238–243. We need to know if this exchange took place before or during the journey to Kashima in 1687.

21. S. Hamill (Tr.). *The Essential Basho* (Boston, MA: Shambhala Publications, 1999), 180.

22. J. Reichhold (Tr.). *Basho. The Complete Haiku* (Tokyo: Kodansha International, 2008), 409.

23. R. Aitken, *A Zen Wave*, 76. Aitken discusses such an issue in a different *haiku*.

24. D. Suzuki, *Zen and Japanese Culture*, 228.

25. J. Austin. Avian Zen. *The Eastern Buddhist. New Series* 2013; 44/1: 121–131.

26. D. Suzuki, *Zen and Japanese Culture*, 239–240.

27. D. Suzuki, ibid., 240. Here, Suzuki might seem to be defining this old-pond *haiku* as "sensation triggers intuition."

28. In the present author's interpretation, such "depths" correspond with known neurobiological facts. From this perspective, they express the covert instinctual capacities of the human brain. Having emerged and been refined during random evolutionary events on this planet, these innate neural capacities need not evoke a supernatural explanation nor represent some kind of exotic "Cosmic Unconscious" state.

29. D. Suzuki, *Zen and Japanese Culture*, 243.

30. D. Suzuki, ibid., 238.

31. M. Ueda, *Basho and His Interpreters*, 138, 140–142.

32. J. Reichhold, *Basho. The Complete Haiku*, 262.

33. J. Reichhold, ibid., 59, 402. She dates this "frog" *haiku* to the spring of 1682 or 1681.

34. J. Reichhold, ibid., 402.

35. N. Bolognini, C. Silvia, F. Martina, et al. The sound-induced phosphene illusion. *Experimental Brain Research* 2003; 231(4): 469–478. doi: 10.1007/s00221-013-3711-1.

36. C. Parkinson, L. Shari, and T. Wheatley. A common cortical metric for spatial, temporal, and social distance. *Journal of Neuroscience* 2014; 34(5): 1979–1987. doi: 10.1523/JNEUROSCI.2159. The latest region representing our normal Self-referent sense of personal "turf" has been identified. It arises on the right side, nearer the

angular gyrus than the adjacent posterior superior temporal gyrus (chapter 15).

37. S. Hamill, *The Essential Basho*, xxxii—xxxiii.
38. D. Barnhill. (tr.) *Basho's Journey. The Literary Prose of Matsuo Basho* (Albany, NY: State University of New York Press, 2005), 53.
39. D. Barnhill, ibid., 53.
40. D. Barnhill, ibid., 150, note 12.

Chapter 23. Basho's States of Consciousness

1. P. Storandt (Tr.). *The Path to Bodhidharma, The Teachings of Shodo Harada-Roshi*, (Ed.) Jane Lago. (Boston, MA: Tuttle, 2000), 56.
2. This sentence is the closing line of Faulkner's Nobel Acceptance Speech in 1950. It was a time when the Cold War posed the threat of atomic catastrophe.
3. R. Aitken. *A Zen Wave. Basho's Haiku and Zen* (Washington, DC: Shoemaker & Hoard, 2003), 34.
4. R. Aitken, ibid., 6.
5. S. Hamill (Tr.). *The Essential Basho*. Boston, MA: Shambhala Publications, 1999), xxx.
6. J. Reichhold (Tr.). *Basho. The Complete Haiku* (Tokyo: Kodansha International, 2008), 157–159, 189, 191.
7. M. Ueda (Ed.). *Basho and His Interpreters. Selected Hokku with Commentary* (Stanford, CA: Stanford University Press, 1991), 143.
8. R. Aitken, *A Zen Wave*, 32–33.
9. J. Reichhold, *Basho. The Complete Haiku*, 90, 252, 405.
10. T. Oseko (Ed.). *Basho's Haiku. Literal Translations for Those Who Wish to Read the Original Japanese Text, with Grammatical Analysis and Explanatory Notes* (Tokyo: Maruzen, 1990). This is not a *haiku* that manifests eternity.
11. D. Barnhill (Tr.). *Basho's Haiku. Selected Poems by Matsuo Basho* (Albany, NY: State University of New York Press, 2004), 209.
12. J. Reichhold, *Basho. The Complete Haiku*, 95, 286–287.
13. N. Foster and J. Shoemaker. *The Roaring Stream. A New Zen Reader* (Hopewell, NJ: Ecco Press, 1996), 104–105.
14. A. Ferguson. *Zen's Chinese Heritage. The Masters and Their Teachings* (Boston, MA: Wisdom Publications, 2000), 196–200.
15. M. Ueda, *Basho and His Interpreters*, 145.
16. T. Oseko, *Basho's Haiku*, 63.
17. T. Oseko, ibid., 67. Butcho's personal cipher (*kao*) was selected for the cover and the title page of P. Kapleau, *The Three Pillars of Zen*.

Teaching, Practice and Enlightenment (Boston, MA: Beacon Press, 1972), iv.

18. J. Reichhold, *Basho. The Complete Haiku*, 97.
19. S. Addiss. *The Art of Haiku. Its History through Poems and Paintings by Japanese Masters* (Boston, MA: Shambhala Publications, 2012), 98.
20. J. Reichhold, *Basho. The Complete Haiku*, 228.
21. T. Oseko, *Basho's Haiku*, 67.
22. D. Barnhill, *Basho's Haiku*, 216.
23. S. Addis, *The Art of Haiku*, 38.
24. D. Barnhill, *Basho's Haiku*, 290.
25. R. Aitken, *A Zen Wave*, 30.
26. S. Addiss, *The Art of Haiku*, 98.
27. J. Reichhold, *Basho. The Complete Haiku*, 228.
28. D. Barnhill (Tr.) *Basho's Journey. The Literary Prose of Matsuo Basho* (Albany, NY. State University of New York Press, 2005), 24–25. This journal entry indicates that Butcho also wrote a poem that night. It speaks to an old Ch'an theme: the never-changing light in the sky (our original Buddha nature). It is only the intervening clouds (our delusions) that prevent us from perceiving this ongoing light.
29. D. Suzuki. *Zen and Japanese Culture. Bollingen Series LXIV* (Princeton, NJ: Princeton University Press, 1973), 224–226.
30. P. Storandt (Tr.). *Moon by the Window. The Calligraphy and Zen Insights of Shodo Harada*. Eds. T. Williams and J. Lago (Boston, MA: Wisdom, 2011), 74–75.

Chapter 24. Zen and the Daily-Life Incremental Training of Basho's Attention

1. S. Sahn. *The Compass of Zen* (Boston, MA: Shambhala Publications, 1997), 234.
2. J. Reichhold. (Tr.). *Basho. The Complete Haiku* (Tokyo: Kodansha International, 2008), 9.
3. Two paintings portray Basho as he appeared to his contemporary artist-disciples. One is by Kyoriku. [see H. Shirane. *Traces of Dreams, Landscape, Cultural Memory, and the Poetry of Basho* (Stanford, CA: Stanford University Press, 1998), 214]. The other is by Ogawa Haritsu [see the frontispiece in M. Ueda (Ed.). *Basho and His Interpreters. Selected Hokku with Commentary* (Stanford, CA: Stanford University Press, 1991]. Basho's own *haiku*-painting (*haiga*) illustrates both his banana/plantain plant and his closed gate.

[See this full-color plate in S. Addiss. *The Art of Haiku. Its History through Poems and Paintings by Japanese Masters* (Boston, MA: Shambhala Publications, 2012), 3–2.].

4. J. Austin. Avian Zen. *The Eastern Buddhist. New Series* 2013; 44/1: 121–131.

5. In Google, if you enter *skylarks*, you can hear a YouTube version of this bird's song and Vaughan Williams' tone poem, titled *The Lark Ascending*.

6. D. Suzuki. *Zen and Japanese Culture*. Bollingen Series LXIV (Princeton, NJ: Princeton University Press, 1959), 224.

7. Reichhold, *Basho. The Complete Haiku*, 188–191.

8. S. Addiss. *The Art of Haiku. Its History through Poems and Paintings by Japanese Masters* (Boston, MA: Shambhala Publications, 2012), 119.

9. L. Bourque and K. Back. Language, Society, and Subjective Experience. *Sociometry* 1971; 34: 1–21. This survey was based on 1553 interviews.

10. D. MacPhillamy. Some Personality Effects of Long-Term Zen Monasticism and Religious Understanding. *Journal for the Scientific Study of Religion* 1986; 26: 304–319.

11. L. Nielson and A. Kaszniak. Awareness of Subtle Emotional Feelings: A Comparison of Long-Term Meditators and Nonmediators. *Emotion* 2006; 6(3): 392–405.

12. Reichhold, *Basho. The Complete Haiku*, 231, 393.

13. S. Hamill (Tr. 1999). *The Essential Basho* (Boston, MA: Shambhala Publications, 1999), 176.

14. A. Ferguson. *Zen's Chinese Heritage. The Masters and Their Teachings* (Boston, MA: Wisdom Publications, 2000), 37–41.

15. V. Sorenson. *Zen Birds* (Cambridge, MN: Adventure, 2010), 3–4.

16. Swallow-embracing space (!) No human presence. These three lines complement the earlier words of Master Hakuin written three centuries earlier: "For all people crossing the ocean of life and death, how enviable is the flight of the swallow." See A. Seo and S. Addiss. *The Sound of One Hand: Painting and Calligraphy by Zen Master Haikuin.* (Boston, MA: Shambhala Publications, 2010), 152.

17. The suggestion of a mixed message does not exclude the possibility that an absorption ushered in *kensho*, as apparently happened in the case of D. T. Suzuki. [ZB: 477]

18. S. Addis, *The Art of Haiku*, 121–126. Addis provides these reasons for Basho's preeminence: (1) He varied and fit no niche, (2) He

engaged novel settings on his travels, (3) He still appreciated the traditions of the past, (4) He was deeply humble, not Self-satisfied, (5) He was curious, nonjudgmental, (6) His humor was subtle, (7) He could combine two different images into one *haiku*, (8) He used brush paintings (*haiga*) to enhance his poetry, (9) He evolved toward *lightness* (*karumi*), (10) His seasons and natural elements expressed human qualities.

Chapter 25. A Story about Wild Birds, Transformed Attitudes, and a Supervisory Self

1. D. Brooks. *The Road to Character* (New York, NY; Random House, 2015), 201–206. The classic story of a dramatic conversion is that of Saul of Tarsus on the road to Damascus (*The New Testament*, Acts, ix, 1–9). The light "shined round about him." Falling to the ground, he heard a voice [Jesus] saying "Saul, Saul why persecutest thou me?"

2. A. Morin. Inner Speech, in *Encyclopedia of Human Behavior* second edition V. Ramachandran (Ed.). (Oxford: Elsevier/Academic Press, 2012), 436–443.

3. B. Alderson-Day, S. Weis, S. McCarthy-Jones, et al. The Brain's Conversation with Itself: Neural Substrates of Dialogic Inner Speech. *Social Cognitive and Affective Neuroscience* 2016; 11(1): 110–120.

4. M. Peroone-Bertolotti, L. Rapin, J.-P. Lachaux, et al. What Is That Little Voice Inside My Head? Inner Speech Phenomenology, Its Role in Cognitive Performance, and Its Relation to Self-Monitoring. *Behavioral Brain Research* 2014; 261: 220–238. In an earlier report from this group, intracerebral electrodes monitored silent reading. A "wavering sequence of activations" occurred in the auditory cortex during this silent reading. See M. Perrone-Bertolotti, J. Kujala, J. Vidal, et al. How Silent Is Silent Reading? Intracerebral Evidence for Top-Down Activation of Temporal Voice Areas During Reading. *Journal of Neuroscience* 2012; 32(49): 17554–17562.

5. C. Wilson-Mendenhall, W. Simmons, A. Martin, et al. Contextual Processing of Abstract Concepts Reveals Neural Representations of Nonlinguistic Semantic Content. *Journal of Cognitive Neuroscience* 2013; 25(6): 920–935.

6. V. Wedeen, D. Rosene, R. Want, et al. The Geometric Structure of the Brain Fiber Pathways. *Science* 2012; 335(6076): 1628–1634.

In Closing

1. N. Foster and J. Shoemaker (Eds.). *The Roaring Stream. A New Zen Reader* (Hopewell, NJ Ecco Press, 1996), 342–350. Ryokan wrote these lines on his deathbed. He was the Japanese equivalent of a poet laureate. Like Basho (pp. 304–313), who lived a century earlier, Ryokan led the legendary life of a hermit on pilgrimage, writing poetry ad lib. Unlike Basho, Ryokan had received formal training in Soto Zen and was authorized to teach it.

2. Robert Aitken-Roshi wrote the foreword to this same volume above (ix–xi). Here, he states the case for Zen students reading books, not avoiding them. It was when Aitken first discovered R. H. Blyth's book about Zen that "new and marvelous vistas of culture and thought opened for me." After Aitken's lucky encounter with Blyth in a Japanese prison camp, his own experience with good books then "put the lie to the commonly heard notion that reading and study are at odds with Zen practice."

Appendix A. Back to Nature: Pausing in Awe

1. P. Piff, P. Dietz, M. Feinberg, et al. Awe, the Small Self, and Prosocial Behavior. *Journal of Personality and Social Psychology* 2015; 108(6): 883–899. The Tasmanian eucalyptus trees in this campus grove are the tallest hardwood trees in North America.

2. J. Stellar, N. John-Henderson, C. Anderson, et al. Positive Affect and Markers of Inflammation: Discrete Positive Emotions Predict Lower Levels of Inflammatory Cytokines. *Emotion* 2015; 15(2): 129–133.

3. I. van der Ham, A. Faber, M. Venselaar, et al. Ecological Validity of Virtual Environments to Assess Human Navigation Ability. *Frontiers in Psychology* 2015; 6: 637.

4. The demand characteristics associated with artificial virtual environments pose additional difficulties. Near-infrared spectroscopy (NIRS) could be used to study, under field conditions, the effects that natural settings have on the brain. See M. Horiuchi, J. Endo, N. Takayama, et al. Impact of Viewing vs. Not Viewing a Real Forest on Psychological and Psychological Responses in the Same Setting. *International Journal of Environmental Research and Public Health* 2014; 11(10): 10883–10901.

Appendix B. Reminders: The Crucial Role of Inhibitory Neurons and Messenger Molecules in Attentional Processing

1. G. Gregoriou, S. Paneri, and P. Sapountzis. Oscillatory Synchrony as a Mechanism of Attention Processing. *Brain Research* 2015; 1626: 165–182.

2. R. Auksztulewicz and K. Friston. Attentional Enhancement of Auditory Mismatch Responses: a DCM/MEG Study. *Cerebral Cortex* 2015; 25(11): 4273–4283.

3. N. Logothetis. Neural-Event-Triggered fMRI of Large-Scale Neural Networks. *Current Opinion in Neurobiology* 2015; 31: 214–222.

4. M. Saggar, A. Zanesco, B. King, et al. Mean-Field Thalamocortical Modeling of Longitudinal EEG Acquired During Intensive Meditation Training. *NeuroImage* 2015; 114: 88–104.

5. R. Mak-McCully, S. Deiss, B. Rosen, et al. Synchronization of Isolated Downstates (K-Complexes) May Be Caused by Cortically Induced Disruption of Thalamic Spindling. *Public Library of Science Computational Biology* 2014; 10(9): e1003855. doi: 10.1371/journal.pcbi.1003855.

6. P. Halasz, R. Bodizs, L. Parrino, et al. Two Features of Sleep Slow Waves: Homeostatic and Reactive Aspects—From Long Term to Instant Sleep Homeostasis. *Sleep Medicine* 2014; 15(10): 1184–1195.

7. J. Pita-Almenar, D. Yu, H. Lu, et al. Mechanisms Underlying Desynchronization of Cholinergic-Evoked Thalamic Network Activity. *Journal of Neuroscience* 2014; 34(43): 14463–14474.

8. L. Lewis, J. Voigts, F. Flores, et al. Thalamic Reticular Nucleus Induces Fast and Local Modulation of Arousal State. *Elife* 2015; Oct 13; 4, pii: e08760. doi: 10.7554/eLife.08760.

9. C. Herrera, M. Cadavieco, S. Jego, et al. Hypothalmic Feedforward Inhibition of Thalamocortical Network Controls Arousal and Consciousness. *Nature Neuroscience* 2016; 19(2): 290–298.

10. R. Brown and J. McKenna. Turning a Negative into A Positive: Ascending GABAergic Control of Cortical Activation and Arousal. *Frontiers in Neurology* 2015; 6: 135. doi: 10.3389/fneur.2015.00135; A. Willis, B. Slater, E. Gribkova, et al. Open-Loop Organization of Thalamic Reticular Nucleus and Dorsal Thalamus: A Computational Model. *Journal of Neurophysiology* 2015; 114(4): 2353–2367. Computational models suggest that some open-loop circuits between the reticular nucleus and thalamocortical nerve cells could enable it to act as a "tunable filter." This means that the reticular nucleus can *enhance* certain thalamocortical activations not simply inhibit them.

11. K. Smigasiewicz, D. Asanowicz, N. Westphal, et al. Bias for the Left Visual Field in Rapid Serial Visual Presentation: Effects of Additional Salient Cues Suggest a Crucial Role of Attention. *Journal of Cognitive Neuroscience* 2015; 27(2): 266–279.

12. S. Irmak, L. de Lecea. Basal Forebrain Cholinergic Modulation of Sleep Transitions. *Sleep* 2014; 37(12): 1941–1951.

13. F. Wilson and E. Rolls. Neuronal Reponses Related to the Novelty and Familiarity of Visual Stimuli in the Substantia Innominata, Diagonal Band of Broca and Periventricular Region of the Primate Basal Forebrain. *Experimental Brain Research* 1990; 80(1): 104–120.

14. G. Esber, K. Torres-Tristani, and P. Holland. Amygdalo-Striatal Interaction in the Enhancement of Stimulus Salience in Associative Learning. *Behavioral Neuroscience* 2015; 129(2): 87–95.

Appendix C. Magnetoencephalograhy

1. L. Liu and A. Ioannides. Emotion Separation Is Completed Early and It Depends on Visual Field Presentation. *Public Library of Science One* 2010; 5(3): e9790.

2. Ibid.

3. W. Luo, T. Holroyd, M. Jones, et al. Neural Dynamics for Facial Threat Processing as Revealed by Gamma Band Synchronization Using MEG. *Neuroimage* 2007; 34(2): 839–847.

4. K. Kveraga, A. Ghuman, and M. Bar. Top-Down Predictions in the Cognitive Brain. *Brain and Cognition* 2007; 65(2): 145–168. Granger analyses of MEG data can infer the *direction* of information flow between two cortical regions. The analysis examines the past history and present data that indicate how activities in the two regions relate to each other. The criteria to establish "connectivity" are met *if* the signal from one region can be predicted based on the past values of its responses to the signal from the other region. The possibility also exists that both cortical regions may also be driven from the same thalamic pacemaker. Directional phase-locking techniques can supplement Granger analyses.

5. A. Ioannides, G. Kostopoulous, L. Liu, et al. MEG Identifies Dorsal Medial Brain Activations During Sleep. *Neuroimage* 2009; 44(2): 455–468.

6. H. Lou, M. Joensson, K. Biermann-Ruben, et al. Recurrent Activity in Higher Order, Modality Non-Specific Brain Regions: a Granger Causality Analysis of Autobiographic Memory Retrieval. *Public Library of Science One* 2011; 6(7): e22286. See also note 4.

7. C. Kerr, S. Jones, Q. Wan, et al. Effects of Mindfulness Meditation Training on Anticipatory Alpha Modulation in Primary Somatosensory Cortex. *Brain Research Bulletin* 2011; 85(3–4): 96–103.

8. C. Styliadis, A. Ioannides, P. Bamidis, et al. Amygdala Responses to Valence and Its Interaction by Arousal Revealed by MEG. *International Journal of Psychophysiology* 2014; 93(1): 121–133. High- and low-arousal pictures were meditated by regions other than the amygdala, such as the anterior cingulate cortex, the temporal pole, and the superior parietal lobule. The five male and five female subjects were self-identified as heterosexual. Their numbers were too few to yield reliable gender conclusions about the 13 subnuclei in the amygdala complex. [SI: 251–252]

Appendix D. Diffusion-Weighted Imaging

1. R. Bammer. Basic Principles of Diffusion-Weighted Imaging. *European Journal of Radiology* 2003; 45(3): 169–184.

2. M. Viallon, V. Cuvinciuc, B. Delattre, et al. State-of-the-Art MRI Techniques in Neuroradiology: Principles, Pitfalls, and Clinical Applications. *Neuroradiology* 2015; 57(5): 441–467. Page 442 considers DWI, DTI, DKI. Other topics reviewed include perfusion imaging, arterial spin labeling, and dynamic contrast enhancements.

3. J. O'Muircheartaigh, S. Keller, G. Barker, et al. White Matter Connectivity of the Thalamus Delineates the Functional Architecture of Competing Thalmocortical Systems. *Cerebral Cortex* 2015; doi: 10.1093/cercor/bhv063. The results of this study do not duplicate the results of fMRI data alone when they are interpreted by independent component analysis (ICA) (as discussed in appendix E).

4. E. Kumral, E. Deveci, C. Erdogan, et al. Isolated Hipocampal Infarcts: Vascular and Neuropsychological Findings. *Journal of Neurological Sciences* 2015; 356(1–2): 83–89; M. Bergui, D. Castagno, F. D'Agata, et al. Selective Vulnerability of Cortical Border Zone to Microembolic Infarct. *Stroke* 2015; 46(7): 1864–1869.

5. D. Osher, R. Saxe, K. Koldewyn, et al. Structural Connectivity Fingerprints Predict Cortical Selectivity for Multiple Visual Categories across Cortex. *Cerebral Cortex* 2015; doi: 10/1093/cercor/bhu303.

6. N. Voets, R. Menke, S. Jbabdi, et al. Thalamo-Cortical Disruption Contributes to Short-Term Memory Deficits in Patients with Medial Temporal Lobe Damage. *Cerebral Cortex* 2015; doi: 10/1093/cercor/bhv109.

7. A. Eden, J. Schreiber, A. Anwander, et al. Emotions Regulation and Trait Anxiety are Predicted by the Microstructure of Fibers between Amygdala and Prefrontal Cortex. *Journal of Neuroscience* 2015; 35(15): 6020–6027. The anxiety data that define the two groups need to be clarified.

8. S. Umesh Rudrapatna, A. Hamming, M. Wermer, et al. Measurements of Distinctive Features of Cortical Spreading Depolarizations with Different MRI Contrasts. *NMR in Biomedicine* 2015; 28(5): 591–600.

9. R. Williams, D. Reutens, and J. Hocking. Functional Location of the Human Color Center by Decreased Water Displacement Using Diffusion-Weighted fMRI. *Brain and Behavior* 2015; 5(11):e00408. doi: 10.1002/brb3.408.

10. T. Bracht, D. Jones, S. Bells, et al. Myelination of the Right Parahippocampal Cingulum is Associated with Physical Activity in Young Healthy Adults. *Brain Structure and Function* 2016; Jan 19. [Epub ahead of print] High Angular Resolution Diffusion Imaging (HARDI) improves the detection and resolution of multiple crossing axonal bundles. These HARDI findings complement the evidence pointing toward dynamic networks that link the retrosplenial region with the medial temporal lobe (chapters 6, 12, 13, 15).

11. F. Deligianni, D. Carmichael, G. Zhang, et al. NODDI and Tensor-Based Microstructural Indices as Predictors of Functional Connectivity. *Public Library of Science One* 2016; doi: 10.1371/journal. pone.0153404; B. Scherrer, O. Afacan, M. Taquet, et al. Accelerated High Spatial Resolution Diffusion-Weighted Imaging. *Informational Processing in Medical Imaging* 2015; 9123: 69–81.

12. C. Rae, L. Hughes, M. Anderson, et al. The Prefrontal Cortex Achieves Inhibitory Control by Facilitating Subcortical Motor Pathway Connectivity. *Journal of Neuroscience* 2015; 35(2): 786–794. In DWI, the mean diffusal density corresponds in white matter with the number of axons and their degree of myelination. In this white matter, lower values for mean diffusivity are interpretable as implying *greater* degrees of structural connectivity.

Appendix E. Some Newer Methods of fMRI Analysis

1. S. Whicher, R. Spiller, and W. Williams (Eds.). *The Early Lectures of Ralph Waldo Emerson*, Volume 1 (Cambridge, MA: Harvard University Press, 1959) 23. Emerson gave this lecture, titled "The Uses

of Natural History" in 1833, a year after he had resigned as pastor from the Second [Unitarian] Church in Boston.

2. W. Cheng, E. Rolls, H. Gu, et al. Autism: Reduced Connectivity between Cortical Areas Involved in Face Expression, Theory of Mind, and the Sense of Self. *Brain* 2015; 138 (Pt 5): 1382–1393. The nucleus reuniens is a key midline nucleus in the limbic thalamus (chapter 8). In this study of 418 autistic subjects, why does the medial thalamus show unexpectedly strong connectivity with the middle temporal gyrus/superior temporal region and the postcentral gyrus? Could this reflect the prenatal stage at which this spectrum of disorders begins and help explain some phenomena of Self-centeredness?

Appendix F. The *Enso* on This Cover

1. A. Seo. *Enso. Zen Circles of Enlightenment* (Boston, MA: Weatherhill, 2007), 14–15.
2. T. Terayama. *Zen Brushwork. Focusing the Mind with Calligraphy and Painting* (Tokyo: Kodansha, 2003).

Appendix G. Word Problems

1. S. Shigematsu. *A Zen Forest. Sayings of the Masters* (New York: Weatherhill, 1981), 3.
2. F. Travis. Transcendental Experiences during Meditation Practices. *Annals of the New York Academy of Science* 2014; 1307: 1–8.
3. J. Nash and A. Newberg. Toward a Unifying Taxonomy and Definition for Meditation. *Frontiers in Psychology* 2013; doi: 10.3389/fpsyg.2013.00806.

Index

Abstract concepts, 48, 144, 211
Acetylcholine, 125, 148, 219, 220
Advita meditation, 239
Aesthetic experiences, 202
Affective limbic system, 15, 66, 104
Agnosia, 130
"Aha!" experiences, 16
Aitken, Robert, 173, 174
Allocentric attention. *See also* Self—other relationships
  anonymity of, 35, 40, 172
  bias, 52
  brain processing regions, 53
  categorical description, 35, 172
  looking up and out, 125, 194, 208
  maps of landmarks, 46, 47
  navigation studies, 52–53
  primate studies of, 41, 55
  turning and, 120
  visual neglect, 49, 51
Alpha modulations, 53, 224
Altar ego, 108–109
Alternate Uses Task, 12
Altruism, 27–28
Alzheimer disease, 66
Amygdala
  of altruists, 27–28
  appropriateness processing, 6
  emotional valence processing, 224
  in hippocampal outflow systems, 66
  novelty processing, 6

representing stress-related network, 92
response to visual stimuli, 222
Amyloid angiopathy, 149
Ancient Greeks, 15
Angular gyrus
  anatomy, 36, 47, 135
  connection to parahippocampal gyrus, 50
  egocentric neglect, 49
  language network, 92
  memory networks, 73
  relation to supramarginal gyrus, 138
  role of, 48, 135–136
  sensory/motor distinctions, 136–137
Anxiety, 228
Appropriateness processing, 5, 6
Aspartate, 149
Associations, 212
Athletes, 155–157
Attention. *See also* Allocentric attention; Egocentric attention
  anatomical systems, 35, 73, 152
  dorsal processing system, 61, 131, 152, 238
  effortless, 157
  external focusing, 157
  fixation, 132
  goal-directed, 57, 103
  inflexibility, 132
  "in the zone," 157
  during meditation, 28
  memory and, 73, 82–83
  over attention to stimuli, 132

Calcarine sulcus, 152
Calpain, 149
Career paths, 111
Caudate nucleus, 6, 9, 10, 82
Cerebral artery infarcts, 48, 50
Cerebral cortex, 216
Character development
  altruism, 27–28
  cultural estimates, 24
  native virtues, 26–27
  perfections, 22
  through Living Zen, 162
Chess games. *See Shogi*
Chio (*haiku* poet), 193
Christian path, 209
Chronic epileptic focus, 227
Cingulate cortex
  anterior, 79, 162
  beta decreases, 39
  BOLD signals, 162
  in default network, 36
  divergent thinking, 7
  in medial entorhinal pathway,
    63
  memory regions, 73, 223
  monitoring procedures, 79
  representing stress-related
    network, 92
  volume in meditators, 74
Cingulate gyrus
  anterior, 107
  appropriateness processing, 6
  auditory focusing task
    activation, 107
  cognitive conflict activation, 15
  game playing strategy and, 10
  in hippocampal outflow
    systems, 66
  inner speech and, 210
  integrative body-mind training
    and, 15

in memory circuits, 84
volume declines in meditators,
  74
Clarity, 163, 202, 208, 211
Clark, R., 225
Claustrum, 112
Clinging areas, 154–155
Coidentity, 20
Collateral sulcus, 138, 142
Colliculus, superior, 133
Compassion, 20, 26, 27, 28, 161,
  163
Comprehension, 15
Concentrative meditation, 12,
  111
Conjunctions, 47, 59, 67, 238
Connectome, 55, 212
Conscience, 77, 209
Conscious processing, 23
Content, 64, 239
Context, 63–64, 239
Convergent thinking, 12, 14
Cortical lesions, 49
Coyote, Peter, 164
Cranial nerve X, 143–144
Creative cognition, 11
Creative problem solving
  appropriateness processing, 6
  conflict detection, 15
  divergent approaches to,
    7–8
  enhancement through
    meditation, 11–12, 14, 17
  mindfulness and, 16–17
  movement tasks, 9–10
  novelty processing, 6
  ordinary *vs.* analytical
    approaches, 16–17
  short-term meditation and,
    15–16
  using matchsticks, 4–7

gamma activity, 39, 223
in hippocampal outflow
systems, 66
memory role, 73, 85, 223
motor network, 92
novelty processing, 6
representing stress-related
network, 92
rostral hippocampal system, 66
Self-relational functions, 45
value-conferring role, 43
voice-sensitive area, 143
"Presence" of mind, 108
Present moments, 71, 155, 158,
161
Presubiculum, 56, 58
Presupplementary motor area
(pre-SMA), 231
Pride, 163
Primary sensory cortex, 224
Primary visual cortex, 228
Primate studies, 41–43, 55,
56–57
Prior conditioning, 28, 106
Probabilistic tractography, 228
Problem solving. *See* Creative
problem solving
Procedural memory, 79, 89, 156
Proprioception, 56, 118, 124
Prosocial behaviors
altruism, 27–28
awe, effects of, 214–215
Psyche, 171, 172, 200
Psychological maturation,
201–202
Pulvinar, 223
Putamen, 15, 51

Qigong, 156
Quickenings, 45, 95, 96, 106, 111,
201

Raindrops, 194–195
Rapid tachistoscopic images, 46
Reality
comprehension, 20
of the external world, 40
frames of reference, 171, 172
Ultimate, 115
Universal, 115
Receptive meditation, 12, 153
Reciprocal hippocampal-
parahippocampal system, 66
Recollection. *See* Memory
Reichhold, Jane, 179, 180, 186
Relative egocentric distance, 48
Relaxation training, 14, 15
Religious conversion, 206,
209–210
Religious experiences, 202
Remindfulness, 76, 77–79, 108
Remote Associations Test (RAT),
15, 17
Repetition, 162
Resting state, 7, 36, 82, 91–92,
103
Restructuring, 5
Reticular nucleus, 125, 218
Retraining, 23, 78, 151, 162, 226
Retreats, 164, 166, 181
Retrosplenial cortex
allocentric processing, 53
anatomical location, 44–45
in default system, 36
head direction cells in, 57
in hippocampal outflow
systems, 66
mapping function, 46, 67
in medial entorhinal pathway,
63
in memory circuits, 84
perception of scenes, 227
Ricard, Matthieu, 27